UNCERTAINTY IN GEOMETRIC COMPUTATIONS

THE KLUWER INTERNATIONAL SERIES
IN ENGINEERING AND COMPUTER SCIENCE

UNCERTAINTY IN GEOMETRIC COMPUTATIONS

edited by

Joab Winkler
Mahesan Niranjan
University of Sheffield, United Kingdom

SPRINGER SCIENCE+BUSINESS MEDIA, LLC

Library of Congress Cataloging-in-Publication Data

Uncertainty in geometric computations / edited by Joab Winkler, Mahesan Niranjan.
   p.cm.—(The Kluwer international series in engineering and computer science; SECS 704)
Includes bibliographical references and index.
ISBN 978-1-4613-5252-5   ISBN 978-1-4615-0813-7 (eBook)
DOI 10.1007/978-1-4615-0813-7
  1. Geometry—Data processing—Congresses. I. Winkler, Joab. II. Niranjan, Mahesan.
III. Series.

QA448.D38 U53 2002
516'.00285—dc21

2002028781

Copyright © 2002 by Springer Science+Business Media New York
Originally published by Kluwer Academic Publishers in 2002
Softcover reprint of the hardcover 1st edition 2002

All rights reserved. No part of this work may be reproduced, stored in a retrieval system, or transmitted in any form or by any means, electronic, mechanical, photocopying, microfilming, recording, or otherwise, without the written permission from the Publisher, with the exception of any material supplied specifically for the purpose of being entered and executed on a computer system, for exclusive use by the purchaser of the work.

Permission for books published in Europe: permissions@wkap.nl
Permission for books published in the United States of America: permissions@wkap.com

Printed on acid-free paper.

Contents

CONTRIBUTORS AND PARTICIPANTS	vii
PREFACE	xvii

1
Affine Intervals in a CSG Geometric Modeller 1
Adrian Bowyer, Ralph Martin, Huahao Shou, Irina Voiculescu

2
Fast and Reliable Plotting of Implicit Curves 15
Katja Buehler

3
Data Assimilation with Sequential Gaussian Processes 29
Lehel Csato, Dan Cornford, Manfred Opper

4
Conformal Geometry, Euclidean Space and Geometric Algebra 41
Chris Doran, Anthony Lasenby, Joan Lasenby

5
Towards the Robust Intersection of Implicit Quadrics 59
Laurent Dupont, Sylvain Lazard, Sylvain Petitjean, Daniel Lazard

6
Computational Geometry and Uncertainty 69
A.R. Forrest

7
Geometric Uncertainty in Solid Modeling 79
Pierre J. Malraison, William A. Denker

8
Reliable Geometric Computations with Algebraic Primitives and Predicates 91
Mark Foskey, Dinesh Manocha, Tim Culver, John Keyser, Shankar Krishnan

9
Feature Localization Error in 3D Computer Vision 107
Daniel D. Morris and *Takeo Kanade*

10
Bayesian Analysis of Computer Model Outputs 119
Jeremy Oakley, Anthony O'Hagan

11
b 131
T. Poggio, S. Mukherjee, R. Rifkin, A. Raklin, A. Verri

12
Affine Arithmetic and Bernstein Hull Methods for Algebraic Curve Drawing 143
Huahao Shou, Ralph Martin, Guojin Wang, Irina Voiculescu, Adrian Bowyer

13
Local Polynomial Metrics for k Nearest Neighbor Classifiers 155
Robert R. Snapp

14
Visualisation of Incomplete Data Using Class Information Constraints 165
Yi Sun, Peter Tino, Ian Nabney

15
Towards Videorealistic Synthetic Visual Speech 175
Barry Theobald, J. Andrew Bangham, Silko Kruse, Gavin Cawley, Iain Matthews

16
Properties of the Companion Matrix Resultant for Bernstein Polynomials 185
Joab R. Winkler

17
Computation with a Number of Neurons 199
Si Wu, Danmei Chen

INDEX 209

Contributors and Participants

SHUN–ICHI AMARI, *Laboratory for Mathematical Neuroscience, Brain Science Institute, RIKEN, Hirosawa 2–1, Wako–shi, Saitama 351–0198, Japan*
[amari@brain.riken.go.jp]

ANDREW BANGHAM, *School of Information Systems, University of East Anglia, Norwich NR4 7TJ, United Kingdom*
[ab@sys.uea.ac.uk]

ADRIAN BOWYER, *Department of Mechanical Engineering, Bath University, Bath BA2 7AY, United Kingdom*
[a.bowyer@bath.ac.uk]

KATJA BUEHLER, *Institute of Computer Graphics and Algorithms, Vienna University of Technology, A–1040 Vienna, Austria*
[katja@cg.tuwien.ac.at]

BERNARD BUXTON, *Department of Computer Science, University College London, Gower Street, London WC1E 6BT, United Kingdom*
[b.buxton@cs.ucl.ac.uk]

GAVIN CAWLEY, *School of Information Systems, University of East Anglia, Norwich NR4 7TJ, United Kingdom*
[gcc@sys.uea.ac.uk]

DANMEI CHEN, *Department of Computer Science, Sheffield University, 211 Portobello Street, Sheffield S1 4DP, United Kingdom*
[d.chen@dcs.shef.ac.uk]

DAN CORNFORD, *Neural Computing Research Group, Aston University, Birmingham B4 7ET, United Kingdom*
[d.cornford@aston.ac.uk]

LEHEL CSATO, *Neural Computing Research Group, Aston University, Birmingham B4 7ET, United Kingdom*
[csatol@aston.ac.uk]

TIM CULVER, *Think3, 365 Boston Post Road, Sudbury, MA 01776, USA*
[culver@acm.org]

WILLIAM DENKER, *7382 Mt. Sherman Road, Longmont, CO 80503, USA*
[bill_denker@yahoo.com]

CHRIS DORAN, *Cavendish Laboratory, Cambridge University, Madingley Road, Cambridge CB3 OHE, United Kingdom*
[c.doran@mrao.cam.ac.uk]

LAURENT DUPONT, *Loria–CNRS, B.P. 239, 54506 Vandoeuvre-les-Nancy cedex, France*
[dupont@loria.fr]

MARK EASTLICK, *Department of Computer Science, Sheffield University, 211 Portobello Street, Sheffield S1 4DP, United Kingdom*
[m.eastlick@dcs.shef.ac.uk]

ALAN EDELMAN, *Department of Mathematics, Massachusetts Institute of Technology, Cambridge, MA 02139, USA*
[edelman@mit.edu]

ANDREW FITZGIBBON, *Department of Engineering Science, Oxford University, Parks Road, Oxford OX1 3PJ, United Kingdom*
[andrew.fitzgibbon@eng.ox.ac.uk]

ROBIN FORREST, *School of Information Systems, University of East Anglia, Norwich NR4 7TJ, United Kingdom*
[forrest@sys.uea.ac.uk]

MARK FOSKEY, *Department of Computer Science, University of North Carolina at Chapel Hill, Chapel Hill, NC 27599–3175, USA*
[foskey@cs.unc.edu]

YOSHI GOTOH, *Department of Computer Science, Sheffield University, 211 Portobello Street, Sheffield S1 4DP, United Kingdom*
[y.gotoh@dcs.shef.ac.uk]

STEVE GUNN, *Department of Electronics and Computer Science, Southampton University, Southampton SO17 1BJ, United Kingdom*
[srg@ecs.soton.ac.uk]

ROB HARRISON, *Department of Automatic Control and Systems Engineering, Sheffield University, Mappin Street, Sheffield S1 3JD, United Kingdom*
[r.f.harrison@sheffield.ac.uk]

ANDREW HEATON, *D–Cubed Ltd., Park House, Castle Park, Cambridge, CB3 0DU, United Kingdom*
[andrew.heaton@d-cubed.co.uk]

NICHOLAS HIGHAM, *Department of Mathematics, Manchester University, Oxford Road, Manchester M13 9PL, United Kingdom*
[higham@ma.man.ac.uk]

ROLAND HOUGS, *School of Information Systems, University of East Anglia, Norwich NR4 7TJ, United Kingdom*
[roland.hougs@uea.ac.uk]

ATA KABAN, *CIS Division, University of Paisley, High Street, Paisley PA1 2BE, United Kingdom*
[kaba-ci0@wpmail.paisley.ac.uk]

VISAKAN KADIRKAMANATHAN, *Department of Automatic Control and Systems Engineering, Sheffield University, Mappin Street, Sheffield S1 3JD, United Kingdom*
[visakan@sheffield.ac.uk]

TAKEO KANADE, *Robotics Institute, Carnegie Mellon University, Pittsburgh, PA 15213, USA*
[tk@ri.cmu.edu]

JASVINDER KANDOLA, *Department of Computer Science, Royal Holloway College, Egham, Surrey TW20 0EX, United Kingdom*
[j.kandola@cs.rhul.ac.uk]

JOHN KEYSER, *Department of Computer Science, Texas A & M University, College Station, TX 77843–3112, USA*
[keyser@cs.tamu.edu]

MASANORI KIMURA, *Department of Mechanical Engineering, Waseda University, 3–4–1 Okubo, Shinjuku–ku, Tokyo 169–8555, Japan*
[kimura@yamaguchi.mech.waseda.ac.jp]

SHANKAR KRISHNAN, *Information and Visualization Research, AT&T Laboratories, Florham Park, NJ 07932–0971, USA*
[krishnas@research.att.com]

SILKO KRUSE, *School of Information Systems, University of East Anglia, Norwich NR4 7TJ, United Kingdom*
[smk@sys.uea.ac.uk]

LUIZA LARSEN, *Computational Imaging Science Group, Radiological Sciences, Guy's Hospital, London SE1 9RT, United Kingdom*
[luiza.larsen@kcl.ac.uk]

ANTHONY LASENBY, *Cavendish Laboratory, Cambridge University, Madingley Road, Cambridge CB3 0HE, United Kingdom*
[a.n.lasenby@mrao.cam.ac.uk]

JOAN LASENBY, *Department of Engineering, Cambridge University, Trumpington Street, Cambridge CB2 1PZ, United Kingdom*
[jl@eng.cam.ac.uk]

NEIL LAWRENCE, *Department of Computer Science, Sheffield University, 211 Portobello Street, Sheffield S1 4DP, United Kingdom*
[n.lawrence@dcs.shef.ac.uk]

DANIEL LAZARD, *LIP6, Universite Pierre et Marie Curie, Boite 168, 4 place Jussieu, 75252, Paris cedex 05, France*
[daniel.lazard@lip6.fr]

SYLVAIN LAZARD, *Loria–CNRS, B.P. 239, 54506 Vandoeuvre-les-Nancy cedex, France*
[lazard@loria.fr]

GUENNADI LIAKHOVETSKI, *Department of Applied Mathematics, Sheffield University, Houndsfield Road, Sheffield S3 7RH, United Kingdom*
[g.liakhovetski@sheffield.ac.uk]

BOJIAN LIANG, *Department of Computer Science, York University, Heslington, York YO10 5DD, United Kingdom*
[bojian@cs.york.ac.uk]

SOFIA AZEREDO LOPES, *Department of Computer Science, Sheffield University, 211 Portobello Street, Sheffield S1 4DP, United Kingdom*
[s.lopes@dcs.shef.ac.uk]

PIERRE MALRAISON, *PlanetCAD, 2520 55th Street, Boulder, CO 80301, USA*
[pierre.malraison@planetcad.com]

DINESH MANOCHA, *Department of Computer Science, University of North Carolina at Chapel Hill, Chapel Hill, NC 27599–3175, USA*
[dm@cs.unc.edu]

GRAEME MANSON, *Department of Mechanical Engineering, Sheffield University, Mappin Street, Sheffield S1 3JD, United Kingdom*
[graeme.manson@sheffield.ac.uk]

RALPH MARTIN, *Department of Computer Science, Cardiff University, PO Box 916, Cardiff CF24 3XF, United Kingdom*
[ralph.martin@cs.cf.ac.uk]

IAIN MATTHEWS, *Robotics Institute, Carnegie Mellon University, Pittsburgh, PA 15213, USA*
[iainm@cs.cmu.edu]

MARTA MILO, *Department of Computer Science, Sheffield University, 211 Portobello Street, Sheffield S1 4DP, United Kingdom*
[m.milo@dcs.shef.ac.uk]

ANDREW MORRIS, *D–Cubed Ltd., Park House, Castle Park, Cambridge CB3 0DU, United Kingdom*
[andrew.morris@d-cubed.co.uk]

DANIEL MORRIS, *Northrop Grumman Corp., 1501 Ardmore Blvd., Pittsburgh, PA 15217, USA*
[daniel_d_morris@mail.northgrum.com]

SAYAN MUKHERJEE, *Center for Biological and Computational Learning, Massachusetts Institute of Technology, 45 Carleton Street, MA 02142, USA*
[sayan@mit.edu]

IAN NABNEY, *Neural Computing Research Group, Aston University, Birmingham, B4 7ET, United Kingdom*
[i.t.nabney@aston.ac.uk]

MAHESAN NIRANJAN, *Department of Computer Science, Sheffield University, 211 Portobello Street, Sheffield S1 4DP, United Kingdom*
[m.niranjan@dcs.shef.ac.uk]

JEREMY OAKLEY, *Department of Probability and Statistics, Sheffield University, Houndsfield Road, Sheffield S3 7RH, United Kingdom*
[j.oakley@sheffield.ac.uk]

ANTHONY O'HAGAN, *Department of Probability and Statistics, Sheffield University, Houndsfield Road, Sheffield S3 7RH, United Kingdom*
[a.ohagan@sheffield.ac.uk]

MANFRED OPPER, *Neural Computing Research Group, Aston University, Birmingham B4 7ET, United Kingdom*
[m.opper@aston.ac.uk]

NICK PEARS, *Department of Computer Science, York University, Heslington, York YO10 5DD, United Kingdom*
[nep@cs.york.ac.uk]

SYLVAIN PETITJEAN, *Loria–CNRS, B.P. 239, 54506 Vandoeuvre-les-Nancy cedex, France*
[petitjea@loria.fr]

DAVID PLATER, *Pathtrace, 45 Boulton Road, Reading RG2 0NH, United Kingdom*
[dplater@pathtrace.com]

TOMASO POGGIO, *Center for Biological and Computational Learning, Massachusetts Institute of Technology, 45 Carleton Street, MA 02142, USA*
[tp@ai.mit.edu]

ALEXANDER RAKHLIN, *Center for Biological and Computational Learning, Massachusetts Institute of Technology, 45 Carleton Street, MA 02142, USA*
[rakhlin@mit.edu]

RYAN RIFKIN, *Center for Biological and Computational Learning, Massachusetts Institute of Technology, 45 Carleton Street, MA 02142, USA*
[rif@mit.edu]

ROMAN ROSIPAL, *NASA Ames Research Center, Mail Stop 269–3, Moffett Field, CA 94035, USA*
[rrosipal@mail.arc.nasa.gov]

ANJALI SAMANI, *Department of Computer Science, University of Sheffield, 211 Portobello Street, Sheffield S1 4DP, United Kingdom*
[a.samani@dcs.shef.ac.uk]

YOSHINAO SHIRAKI, *NTT Communication Science Laboratories, Speech and Motor Control Research Group, 3–1 Morinosato–Wakamiya, Atsugi-shi, Kanagawa 243–0198, Japan*
[shira@idea.brl.ntt.co.jp]

HUAHAO SHOU, *Department of Computer Science, Cardiff University, PO Box 916, Cardiff CF24 3XF, United Kingdom*
[h.shou@cs.cf.ac.uk]

ROD SINGLES, *D–Cubed Ltd., Park House, Castle Park, Cambridge CB3 0DU, United Kingdom*
[rod.singles@d-cubed.co.uk]

ROBERT SNAPP, *Department of Computer Science, University of Vermont, Burlington, VT 05405, USA*
[snapp@cs.uvm.edu]

HANS STETTER, *Am Modenapark 13/4, A–1030 Vienna, Austria*
[stetter@aurora.anum.tuwien.ac.at]

YI SUN, *Neural Computing Research Group, Aston University, Birmingham B4 7ET, United Kingdom*
[suny@aston.ac.uk]

BARRY THEOBALD, *School of Information Systems, University of East Anglia, Norwich NR4 7TJ, United Kingdom*
[b.theobald@uea.ac.uk]

PETER TINO, *Neural Computing Research Group, Aston University, Birmingham, B4 7ET, United Kingdom*
[p.tino@aston.ac.uk]

JULIAN TODD, *NC Graphics Ltd., Silverwood Lodge, Ely Road, Waterbeach, Cambridge CB5 9NN, United Kingdom*
[julian@ncgraphics.net]

EDWARD TUKE, *Department of Computer Science, York University, Heslington, York YO10 5DD, United Kingdom*
[charles.tuke@cs.york.ac.uk]

CAROLE TWINING, *Imaging Science and Biomedical Engineering, University of Manchester, Oxford Road, Manchester M13 9PL, United Kingdom*
[carole.twining@man.ac.uk]

ALESSANDRO VERRI, *DISI, Universita di Genova, via Dodecaneso 35, 16146 Genova, Italy*
[verri@disi.unige.it]

IRINA VOICULESCU, *Computing Laboratory, Oxford University, Parks Road, Oxford OX1 3QD, United Kingdom*
[irina.voiculescu@comlab.ox.ac.uk]

VINCENT WAN, *Department of Computer Science, Sheffield University, 211 Portobello Street, Sheffield S1 4DP, United Kingdom*
[v.wan@dcs.shef.ac.uk]

GUOJIN WANG, *Department of Mathematics, Zhejiang University, Hangzhou, China*
[wgj@math.zju.edu.cn]

ADAM WILMER, *Department of Electronics and Computer Science, Southampton University, Southampton SO17 1BJ, United Kingdom*
[aiw99r@ecs.soton.ac.uk]

JOAB WINKLER, *Department of Computer Science, Sheffield University, 211 Portobello Street, Sheffield S1 4DP, United Kingdom*
[j.winkler@dcs.shef.ac.uk]

SUSANNA WRETH, *D–Cubed Ltd., Park House, Castle Park, Cambridge CB3 0DU, United Kingdom*
[susanna.wreth@d-cubed.co.uk]

SI WU, *Department of Computer Science, Sheffield University, 211 Portobello Street, Sheffield S1 4DP, United Kingdom*
[s.wu@dcs.shef.ac.uk]

FUJIO YAMAGUCHI, *Department of Mechanical Engineering, Waseda University, 3–4–1 Okubo, Shinjuku–ku, Tokyo 169–8555, Japan*
[fujio@mn.waseda.ac.jp]

Preface

This book contains the proceedings of the workshop *Uncertainty in Geometric Computations* that was held in Sheffield, England, July 5-6, 2001. A total of 59 delegates from 5 countries in Europe, North America and Asia attended the workshop. The workshop provided a forum for the discussion of computational methods for quantifying, representing and assessing the effects of uncertainty in geometric computations. It was organised around lectures by invited speakers, and presentations in poster form from participants.

Computer simulations and modelling are used frequently in science and engineering, in applications ranging from the understanding of natural and artificial phenomena, to the design, test and manufacturing stages of production. This widespread use necessarily implies that detailed knowledge of the limitations of computer simulations is required. In particular, the usefulness of a computer simulation is directly dependent on the user's knowledge of the uncertainty in the simulation. Although an understanding of the phenomena being modelled is an important requirement of a good computer simulation, the model will be plagued by deficiencies if the errors and uncertainties in it are not considered when the results are analysed. The applications of computer modelling are large and diverse, but the workshop focussed on the management of uncertainty in three areas : Geometric modelling, computer vision, and computer graphics.

It was hoped that the workshop would provide a challenging environment for postgraduate students and research assistants who want to pursue new research activities after several years in a chosen specialisation. The organisers hope that the wide diversity of topics that were discussed – from support vector machines to resultants – provided a fertile forum for the achievement of this objective.

The printed text contains only black and white diagrams and pictures, but the original versions of some of them are in colour. These colour versions are on the CD that is enclosed with these proceedings.

We wish to thank the Engineering and Physical Sciences Research Council (EPSRC) and London Mathematical Society (LMS) for their financial support of the workshop. We also wish to thank Lance Wobus and Susan Lagerstrom-Fife of Kluwer Academic Publishers for their hard work in ensuring that these proceedings are published as soon as possible after the workshop. The time and help of all the referees who read the papers and improved their content is gratefully acknowledged. Finally, we wish to thank Linda Perna of The Department of Computer Science at The University of Sheffield for her administrative skills during the organisation of the workshop.

Sheffield
April 2002

Joab Winkler
Mahesan Niranjan

Chapter 1

AFFINE INTERVALS IN A CSG GEOMETRIC MODELLER

Adrian Bowyer
Bath University, Bath BA2 7AY, United Kingdom
A.Bowyer@bath.ac.uk

Ralph Martin, Huahao Shou
Cardiff University, Cardiff CF24 3XF, United Kingdom
Ralph.Martin@cs.cf.ac.uk, H.Shou@cs.cf.ac.uk

Irina Voiculescu
Oxford University, Oxford, OX1 3QD, United Kingdom
Irina.Voiculescu@comlab.ox.ac.uk

Abstract Our CSG modeller, svLIs, uses interval arithmetic to categorize implicit functions representing primitive shapes against boxes; this allows an efficient implementation of recursive spatial division to localize the primitives for a variety of purposes, such as rendering or the computation of integral properties.

Affine arithmetic allows a track to be kept on the contributing terms to an interval, which often reduces the conservativeness of interval arithmetic. In particular, by tracking the asymmetric contributions of even and odd powers of intervals that contain zero, tighter bounds can be kept on resulting interval values.

This paper shows how such techniques can be implemented in the svLIs modeller, and offers a comparison of doing so with using conventional interval arithmetic.

Keywords: Function zeros, root finding, interval arithmetic, affine arithmetic, geometric modelling, CSG modelling, set-theoretic modelling

1. Introduction

The svLis CSG geometric modeller [1] represents solids using implicit inequalities at the leaves of a set-theoretic operator tree. Because virtually any implicit function can be used as a leaf, it is extremely versatile in the shapes it can store. In this paper we restrict our attention to implicit polynomial functions.

To evaluate models, we need to know where the zero-surfaces of the implicit equalities that form the boundaries of the inequalities are, as parts of these represent the surface of the object being modelled. To find the zeros of implicit functions, while avoiding missing any parts, svLis takes an axis-aligned box-shaped region of interest and recursively divides it, keeping any sub-boxes that contain parts of the surfaces. The set-theoretic expressions in these sub-boxes are pruned as division proceeds; the terminating condition for the recursion is either that a box is too small, or (much more usually) that pruning has made its contents sufficiently simple to evaluate.

The problem is thus one of classifying an implicit function against a box: deciding whether or not the function has zeros anywhere in the box. To do so, we use various forms of interval arithmetic, and various different equivalent expressions for the functions; for a fuller description of how this is done see [7], [8].

This paper concerns the use of *Affine Arithmetic* [3] to improve box classification: how to do so, and how its performance compares with ordinary interval arithmetic.

2. Affine arithmetic

2.1 Ordinary intervals

Ordinary interval arithmetic [2], [4], [5] represents uncertainty about a value as a range of possible real numbers that it might take. Thus an interval $[2, 6]$ means that a given number lies between 2 and 6. We can then do arithmetic with intervals to give an interval result.

Now, we can represent the boxes that the modeller uses as three intervals in Cartesian coordinates and substitute these intervals into the implicit expressions. This will give an interval which is a conservative estimate of the range of values that the function takes in the box. Thus if the interval is all positive or all negative we definitely know that there is no surface in the box, but if it straddles zero there may or may not be surface.

2.2 Affine intervals

Affine intervals are expressed as a sum of independent unknown terms:

$$a = a_0 + a_1.e_1 + a_2.e_2 + \cdots + a_n.e_n. \qquad (1.1)$$

Here the interval a has n independent sources of uncertainty, represented as $a_1 \ldots a_n$ multiplied by the e_i terms, which each equal $[-1, 1]$. These are all added to a constant locating term, a_0. The interval $[2, 6]$ is easily represented in this way:

$$[2, 6] = 4 + 2.e_1 = 4 + 2.[-1, 1]. \tag{1.2}$$

If we do arithmetic with such affine intervals we keep them in the form in Expression 1.1, and we can usually get a result with tighter bounds than when using ordinary interval arithmetic, as the independent sources of uncertainty are retained and sometimes cancel. As a simple example, consider evaluating

$$[2, 6] + [2, 6] - [2, 6] \tag{1.3}$$

using ordinary interval arithmetic; this gives $[-2, 10]$. But if they really are all the same interval then affine arithmetic gives

$$(4+2.e_1) + (4+2.e_1) - (4+2.e_1) = 4+2.e_1 = 4+2.[-1, 1] = [2, 6]. \tag{1.4}$$

If they are not the same interval we get

$$(4 + 2.e_1) + (4 + 2.e_2) - (4 + 2.e_3) = [-2, 10] \tag{1.5}$$

the same answer as ordinary interval arithmetic [4].

Affine arithmetic does not always do better than ordinary intervals. Consider that

$$[1, 5]^2 = [1, 25] \tag{1.6}$$

with ordinary interval arithmetic. But

$$(3 + 2.e_1)^2 = 9 + 12.e_1 + 4.e_1^2 = 9 + [-12, 12] + [0, 4] = [-3, 25]. \tag{1.7}$$

An operation which must give a wholly positive result produces an interval straddling zero, and further has produced a wider one than ordinary interval arithmetic. We may be able to alleviate this problem by recognizing, when evaluating the affine expression, that in this case repeated e_1 terms refer to the same thing—more on this below. Note in passing that when we raise one of our e_i terms to an even power the result is a term half as wide as the original: $[0, 1]$ instead of $[-1, 1]$.

Another problem with affine intervals sometimes occurs when one multiplies two independent ones. For example, consider $[0, 2] \times [3, 7]$. With ordinary interval arithmetic the result is $[0, 14]$, the tightest possible result. But using affine arithmetic one gets:

$$(1 + e_1) \times (5 + 2e_2) = 5 + 5e_1 + 2e_2 + 2e_1.e_2 = [-4, 14] \tag{1.8}$$

which is not as good. Note here that $e_1.e_2$ is $[-1, 1] \times [-1, 1]$ which is $[-1, 1]$ and is thus different to $[-1, 1]^2$ which is $[0, 1]$.

The cancelling effect will sometimes outweigh the effects just described, and at others it will not; this depends, in part, on the form of the original expression being evaluated. We have reported elsewhere [6] [7] [8] that on balance affine arithmetic usually gives tighter bounds of uncertainty when finding the zeros of functions of two independent variables. Here we consider its operation in a real geometric modeller working in three dimensions, and also consider the costs as well as the benefits of applying it.

2.3 Affine intervals in svLis

SvLis is written in C++, and has had an `interval` class to do conventional interval arithmetic since its inception. Modifying svLis to use affine intervals just involved the re-writing of this class.

We are evaluating implicit functions in boxes, so there are three independent uncertainties in any implicit evaluation. Boxes are represented by

$$x = x_0 + x_1.e_x, \quad y = y_0 + y_1.e_y, \quad \text{and} \quad z = z_0 + z_1.e_z. \tag{1.9}$$

When such a box is substituted into an implicit function $f(x, y, z)$ the result is an affine expression in the form

$$f = a_0 + a_1.e_x + a_2.e_y + a_3.e_z + a_4.e_x^2 + \cdots + a_n.e_x^i.e_y^j.e_z^k. \tag{1.10}$$

Any product of the e values containing one or more odd powers will be $[-1, 1]$; any in which all the powers are even will be $[0, 1]$. Evaluating the affine intervals is just a question of running along the expression adding it up using conventional interval arithmetic (but see below).

3. Experiments and results

The CSG set theory in svLis has no effect on, and is not affected by, interval evaluation; all the information it needs is whether a given leaf implicit function in a box is definitely all positive or all negative, or if it straddles zero and therefore may contribute surface to the box. Thus, experiments were conducted on single implicit functions in boxes. To start with, a simple slightly off-origin sphere

$$f(x, y, z) = (x - 0.1)^2 + (y - 0.2)^2 + (z - 0.3)^2 - 6^2, \tag{1.11}$$

multiplied out to give

$$f(x, y, z) = x^2 - 0.2x + 0.01 + y^2 - 0.4y + 0.04 + z^2 - 0.6z + 0.09 - 36, \tag{1.12}$$

was divided in the box $[-10, 10] \times [-10, 10] \times [-10, 10]$. The sphere was offset to prevent the faces of the division boxes exactly coinciding with its centre, to avoid having a special case.

As expected, both conventional and affine interval arithmetic give similar results: interval division generates 1246 potentially surface-containing boxes, whereas affine generates only 1210. Figure 1.1 shows the leaf boxes in each case. The recursion divided the boxes in half along their longest dimension

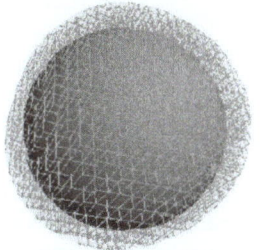 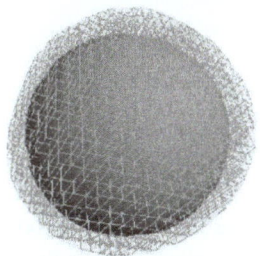

Figure 1.1. Affine (left) and conventional (right) interval division of a sphere to a resolution of 0.0001 times the volume of the original box.

down to a volume of 0.0001 of the initial box where the surface was detected; larger boxes where no surface was detected were left as leaves. The figure shows the surface and the small division boxes.

As seen in Equation 1.3, conventional interval arithmetic suffers if spurious cancelling terms are present in a function being evaluated. These can be removed algebraically before evaluation, but naive users and some automatic processes may leave them extant. To see their effect, we added and subtracted x, y and z terms to Equation 1.12 leading to a new equation with a series of $x + y + z$ terms at the front and a corresponding series of $-x - y - z$ terms at the end. Figure 1.2 shows the results. This is a particularly obvious set of cancelling terms, but it is reasonable to expect similar or worse behaviour for more subtle cancelling algebra when using ordinary intervals.

The work done in recursive division to find the surface of the sphere and other functions is clearly greater if the surface area is greater. Consequently all measures given in this paper are normalized by dividing them by the area of the surface concerned to give work-per-unit-area. The surface area of the sphere was 452.4 square units.

As the leaf boxes are a constant size (0.0001 times the original box volume), each contributes equally to the volume of uncertainty surrounding the surface. Figure 1.3 shows how the number of boxes per unit area of the surface of the sphere (and hence the volume of uncertainty) increases with the addition of spurious cancelling terms (such as $+x \ldots -x$) into Equation 1.12 with interval arithmetic. Affine arithmetic is clearly unaffected, and the number of boxes

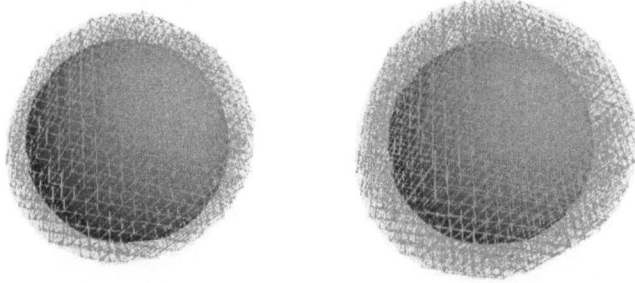

Figure 1.2. Affine (left) and conventional (right) interval division of the sphere to a resolution of 0.0001 times the volume of the original box with 12 extra cancelling terms in its expression. The affine result is identical to the left hand picture in Figure 1.1.

per unit area stays constant at 2.675. The price that is paid for the much better

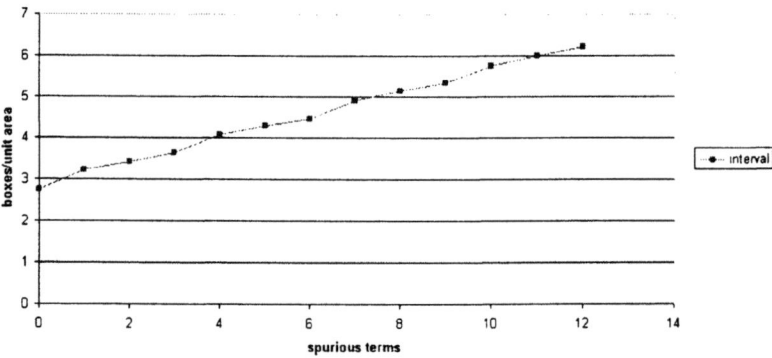

Figure 1.3. A plot of the increase in box count per unit area for conventional interval arithmetic being used to divide the sphere.

performance of affine arithmetic in this case is increased execution time (times are in seconds on a 400MHz Pentium II running Linux). See Figure 1.4.

Two main factors in the cost of testing implicit functions against boxes are the polynomial degree and the fineness of the recursive division. For the main experiments reported in this paper we used polynomials of degrees 1 to 5, and a range of division resolutions from 10^{-1} to 10^{-6} of the volume of the original box. Different applications will require answers to different accuracies. The

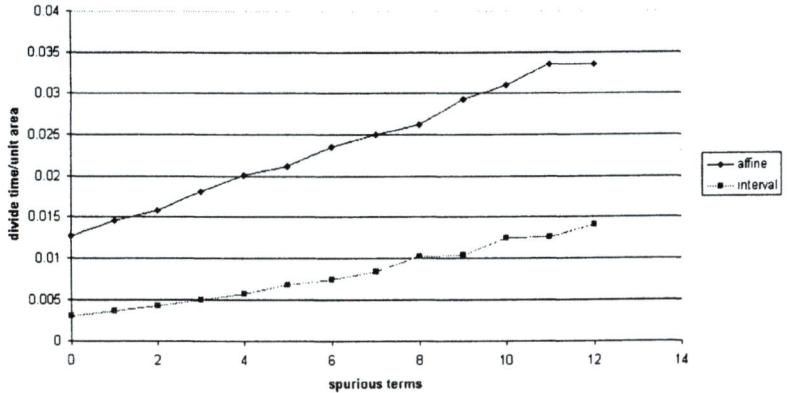

Figure 1.4. A plot of the increase in divide time per unit area for both affine and conventional interval arithmetic being used to divide the sphere.

five functions were:

$$f_1(x, y, z) = 0.557086x - 0.371391y + 0.742781z - 0.204265, \quad (1.13)$$

$$f_2(x, y, z) = 0.06(x^2 - x + 4y - xy + 2yz + 3), \quad (1.14)$$

$$f_3(x, y, z) = 0.06(x^3 - 2x^2 - 5x - 2xy + 4yz + 6), \quad (1.15)$$

$$f_4(x, y, z) = 0.06(0.1x^4 - 2x^2y - 5x^2 - 2xy + 4yz + 6x), \quad (1.16)$$

$$f_5(x, y, z) = 0.06(0.1x^4 - 0.05x^2yz^2 - 5x^2 - 2xy + 4yz + 6x). \quad (1.17)$$

See Figure 1.5. They were chosen to be reasonably similar in their number of terms, and to include some squared and cross-multiple terms (for the reasons given in Equations 1.7 and 1.8 above).

Good approximations to the areas in square units of the five surfaces in that $[-10, 10] \times [-10, 10] \times [-10, 10]$ box were computed using the triangulations generated by svLIs to produce the figures; the areas of the triangles were added up. The results were:

degree	1	2	3	4	5
area	516	619	621	804	1039

First each function was divided down to a leaf box volume of 0.0001 times the original box. Figure 1.6 shows the results for the degree 2 surface, Figure 1.7 shows them for degree 5. Figure 1.8 shows how the number of boxes per unit area of surface changes with polynomial degree.

Unsurprisingly, the linear function behaves identically with both affine and ordinary intervals, but all other functions behave significantly better with affine

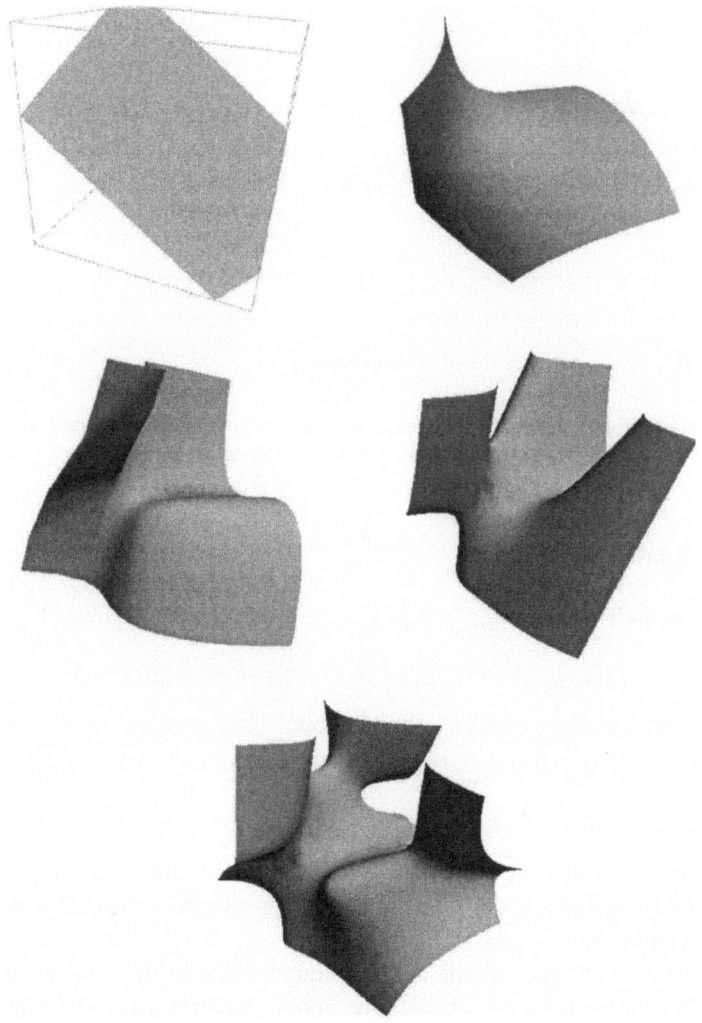

Figure 1.5. The five test functions in the box $[-10, 10] \times [-10, 10] \times [-10, 10]$. The box is shown with the degree 1 surface. The viewpoints are not all the same—they were chosen individually to make the surfaces clear; but in all subsequent pictures of each surface it is shown from the viewpoint that it is shown from here.

arithmetic. As the division resolution was kept constant at 0.0001, the volume of uncertainty depends entirely on the number of boxes per unit area, and so has not been plotted as well. Figure 1.9 shows how the timings compare; as

Affine Intervals in a CSG Geometric Modeller 9

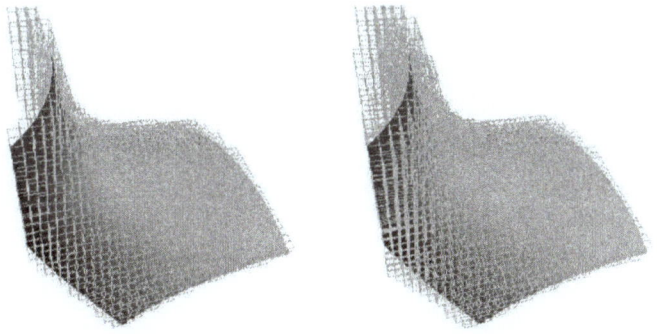

Figure 1.6. The division of the degree 2 surface to a resolution of 0.0001 of the volume of the original box. The affine result is on the left, and the conventional interval result on the right. As SVLIS is intended for engineering modelling it culls backward-pointing faces for efficiency when rendering, so there appears to be no surface in the top-left of the pictures.

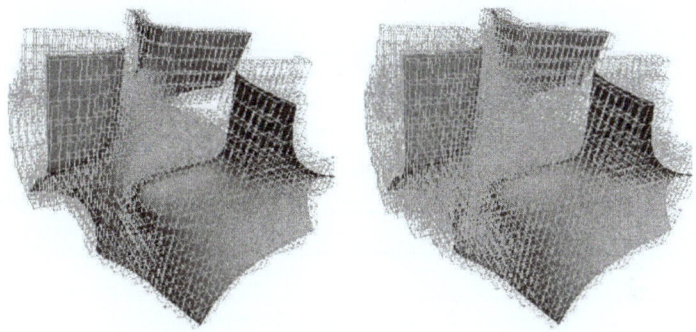

Figure 1.7. The division of the degree 5 surface to a resolution of 0.0001 of the volume of the original box. The affine result is on the left, and the conventional interval result on the right.

would be expected, affine arithmetic is more expensive, and gets worse as the degree increases and more computations have to be done.

In the next experiment, the degree 3 surface was divided to a range of resolutions. Figure 1.10 shows the results for a resolution of 10^{-2}, and Figure 1.11 shows them for 10^{-6}. Figure 1.12 shows how the number of boxes generated depends on resolution, and Figure 1.13 shows how the volume of uncertainty changes. The left hand ends of these graphs are dominated by the fact that the leaf boxes are quite big compared to the original box. Once they get small, the box count settles down so that affine division gives about half the number

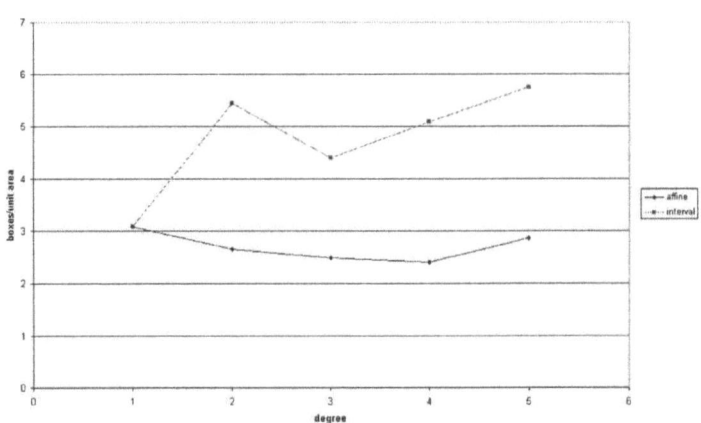

Figure 1.8. The dependence of the number of division boxes generated on degree.

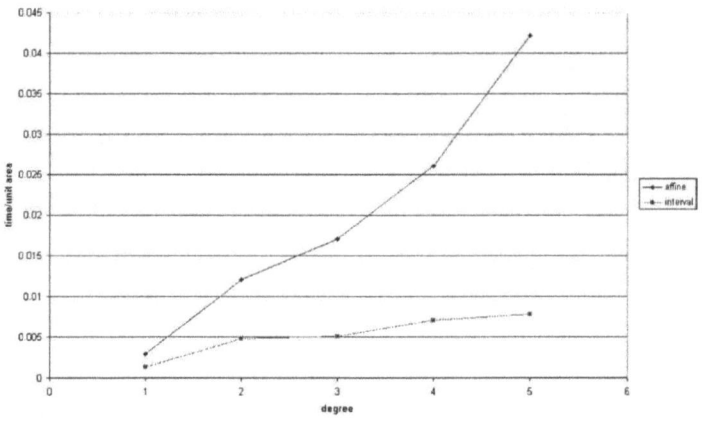

Figure 1.9. The dependence of the time of division on degree.

of boxes that conventional intervals do for any given resolution. Similarly, the volume of uncertainty is about 1.7 times better for affine arithmetic. Finally, Figure 1.14 shows how division times depend on resolution. Here we can see that affine arithmetic is about 3 times slower than interval arithmetic at any resolution. An alternative way of putting this is that one may divide to a leaf box volume about 10 times finer using interval arithmetic in the time that affine arithmetic takes. Looking at Figure 1.13 it can be seen that this would give roughly the same volume of uncertainty, so on that measure, the two methods

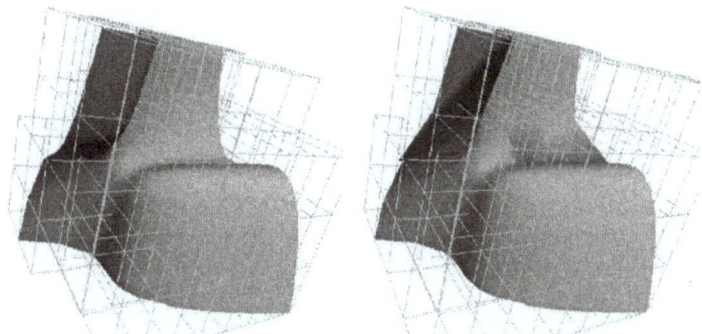

Figure 1.10. The degree 3 surface divided to a resolution of 10^{-2}, showing the affine division on the left and the ordinary interval division on the right.

Figure 1.11. The degree 3 surface divided to a resolution of 10^{-6}, showing the affine division on the left and the ordinary interval division on the right.

are very similar—one gets the same benefit for the same cost. But of course, conventional interval arithmetic has had to generate many more smaller boxes to achieve this, leading to extra storage requirements, and more work when the resulting larger tree of boxes is traversed. Depending on the application, the traversal may need to be done many times once a division is complete, and that extra cost would be incurred with each traversal.

4. Conclusions

We conclude that (i) a few very simple functions, such as x^2, sometimes give tighter bounds using conventional intervals than they do with affine intervals; (ii) in almost all other circumstances affine arithmetic works more accurately than conventional interval arithmetic; (iii) the computational costs of

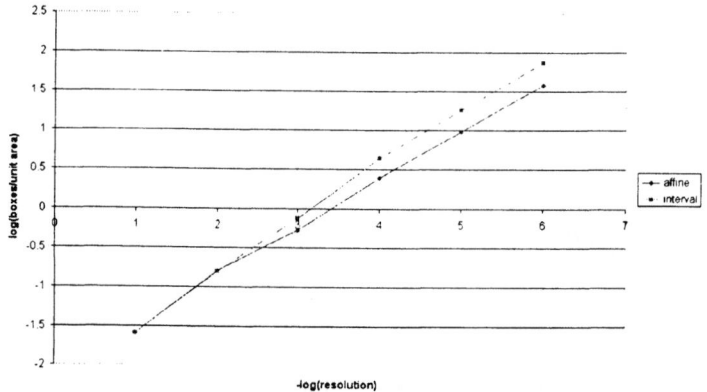

Figure 1.12. How the number of boxes generated depends on division resolution.

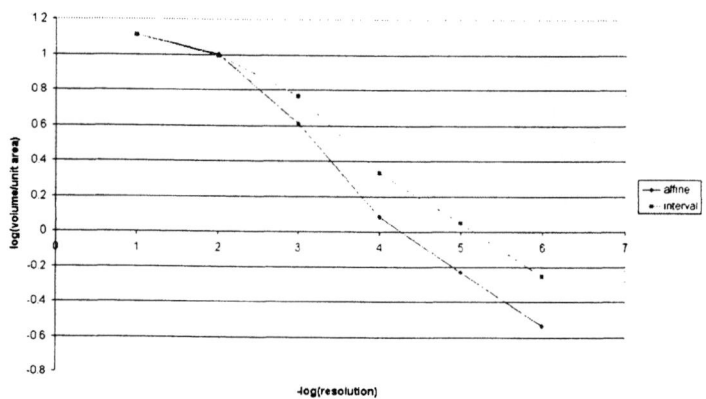

Figure 1.13. How the volume of uncertainty depends on division resolution.

affine arithmetic in our implementation are about 4 times higher for performing any given operation than those for conventional interval arithmetic; (iv) if one wishes to minimize the volume of uncertainty around a surface, there is little to choose between the two methods; (v) if one wishes to minimize the number of surface-containing boxes around a surface and the consequent size of the recursive division tree, affine arithmetic is about twice as good as conventional intervals; (vi) the previous two results mean that affine intervals are almost always at least as good as conventional intervals, and in the vast majority of cases are much better.

Affine Intervals in a CSG Geometric Modeller 13

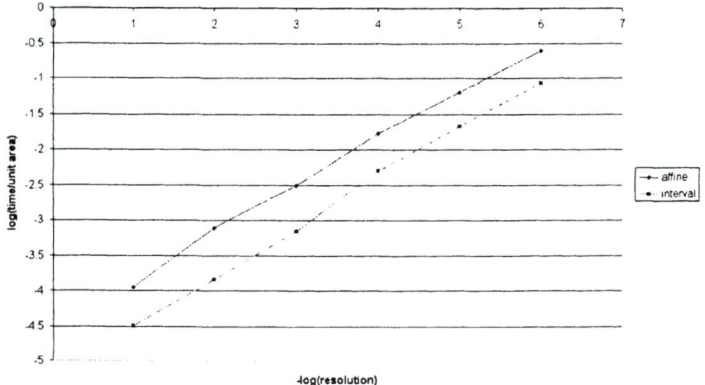

Figure 1.14. How the division time depends on division resolution.

5. A problem for the future

The first conclusion above surprised us initially. Consider the squaring problem discussed earlier:

$$[1,5]^2 \neq (3 + 2.e_1)^2 = 9 + 12.e_1 + 4.e_1^2 = 9 + [-12, 12] + [0, 4] = [-3, 25] \tag{1.18}$$

If, instead of evaluating it using interval arithmetic, we evaluate it for the extreme ends of e_1, that is -1 and 1, we get:

$$(3 + 2(-1))^2 = 9 + 12(-1) + 4(-1)^2 = 1 \tag{1.19}$$

and

$$(3 + 2(1))^2 = 9 + 12(1) + 4(1)^2 = 25. \tag{1.20}$$

In this case, just working with the ends has given us the tightest possible answer, $[1, 25]$. The reason this works better than interval evaluation is that it uses the extra information given by the repeated occurrence of e_1 in the expanded square.

This clearly does not work in general. For example

$$[-1, 5]^2 \to (2 + 3.e_1)^2 = 4 + 12.e_1 + 9.e_1^2. \tag{1.21}$$

Evaluating this at $e_1 = -1$ and $e_1 = 1$ also gives $[1, 5]$, whereas $[-1, 5]^2 = [0, 25]$ so a part—$[0, 1]$—of the correct result is missing.

However, in general, if we have

$$(a + b.e_1)^2 = a^2 + 2ab.e_1 + b^2.e_1^2, \tag{1.22}$$

then, if $-a/b \in e_1$, we need to extend the interval to 0 after evaluating it using the ends of e_1. This gives the right answer.

In more complicated affine expressions this gets much more difficult, of course. We intend to investigate this further, as it may give a method to take additional advantage of repeated e_i terms in affine expressions to tighten the bounds of the result.

Acknowledgments

We would like to thank the China Scholarship Council for funding H. Shou's one-year visit to the Computer Science Department at Cardiff. We would also like to thank Bath University and the ORS award scheme for funding I. Voiculescu for her three years at Bath prior to her current appointment at Oxford.

References

[1] A. Bowyer: *SvLis: set–theoretic kernel modeller*, Information Geometers Ltd, 1995, ISBN 1-874728-07-0 and
http://www.bath.ac.uk/ ~ensab/G_mod/Svlis/.

[2] A. Bowyer, J. Berchtold, D. Eisenthal, I. Voiculescu, K. Wise: *Interval methods in geometric modelling*, in Geometric Modelling and Processing 2000, Eds R. Martin and W. Wang, IEEE Computer Society Press, 2000, pp 321–327, ISBN 0-7695-0562-7.

[3] J.L.D. Comba, J. Stolfi: *Affine arithmetic and its applications to computer graphics*, Proceedings of Anais do VII SIBGRAPI (Brazilian Symposium on Computer Graphics and Image Processing), Recife, Brazil, 1993, pp 9–18.
See also http:// graphics.stanford.edu/ ~comba/.

[4] R.E. Moore, *Interval Analysis*, Prentice Hall, 1966.

[5] H. Ratschek, J. Rokne: *New Computer Methods for Global Optimization* Ellis Horwood, 1988, ISBN 0-7458-0139-0.

[6] H. Shou, R. Martin, I. Voiculescu, A. Bowyer, G. Wang: *Affine arithmetic in matrix form for polynomial evaluation and algebraic curve drawing*, Progress in Natural Science, 12 (1) 77–81, 2002.

[7] I. Voiculescu: *Implicit function algebra in set-theoretic modelling*, PhD thesis , University of Bath, 2001.

[8] I. Voiculescu, J. Berchtold, A. Bowyer, R. R. Martin, Q. Zhang: *Bernstein Form Polynomials and Affine Arithmetic for Surface Location*, Maths of Surfaces IX, Eds. R. Cipolla & R. R. Martin, Springer, 2000, ISBN 1-85233-358-8.

Chapter 2

FAST AND RELIABLE PLOTTING OF IMPLICIT CURVES

Katja Buehler
Institute of Computer Graphics and Algorithms
Vienna University of Technology, Austria
katja@cg.tuwien.ac.at

Abstract   This paper presents a new, fast and reliable subdivision algorithm for adaptive enumeration and plotting of implicit curves. For this purpose, Implicit Linear Interval Estimations (ILIEs) based on affine arithmetics are introduced. They allow a significant acceleration of the subdivision process and a generation of reliable piecewise linear enclosures for the curve. The algorithm has been tested for algebraic curves of high degree and non-trivial trigonometric curves with remarkable results.

Keywords: Implicit curves, algebraic curves, interval arithmetic, affine arithmetic, rendering, linear interval estimations

1.   Introduction

An implicit defined object in $\mathbf{E}^n$ is given by an equation of the form

$$f(x) = 0, \quad x \in \mathbf{R}^n$$

where f can be either a polynomial or any other real valued function.

Implicitly defined curves and surfaces in 2- and 3-space resp., are a powerful tool in computer graphics. They build e.g. the basis for constructive solid geometry and computer aided design systems, and allow the compact description of complex shapes. The implicit representation has the advantage that it allows a fast test whether a point lies inside ($f(x) > 0$), outside ($f(x) < 0$) or on ($f(x) = 0$) the object. Although the location of the object for visualisation [16, 19, 20, 18, 21, 8] or collision detection [8] is an old problem, finding *fast and reliable* solutions is still a topic of recent research in computer graphics.

Pure visualisation algorithms like the marching square/cubes algorithm, ray casting or the application of particle systems suffer from uncertainties produced by the discrete approach and significant features may be missed. *Enumeration algorithms* for implicit objects locate within a certain precision those areas in space which may contain parts of the object. Enumeration of implicit objects is an often used preprocessing step for polygonisation, collision detection, and ray tracing algorithms.

A simple reliable solution for the enumeration problem is based on recursive *uniform space subdivision*, and an incidence test of an axes aligned box and the curve/surface. Some algorithms subdivide until pixel/voxel size is reached [23, 15, 8, 4], and others follow a hybrid approach and compute for each detected cell a linear approximation of the object.

Adaptive space subdivision techniques take the curvature of the object during the subdivision process into account, followed by a linearisation, as in the hybrid uniform case. Adaptive techniques are faster, but cracks could appear in the linear/polygonal approximation for curves/surfaces due to different levels of subdivision of neighbouring cells. Bloomenthal [3] solved the problem for cracks caused by neighbouring cells differing one level of depth in subdivision. Recently Balsys and Suffern [2] presented an adaptive algorithm solving the problem for an arbitrary level of subdivision.

The heart of each subdivision algorithm is the *reliable test* if the actual box to be examined hits the object or not. The reliability of the incidence test is an important requirement, because it is the condition for the reliability of further computations. Taubin [19] proposes an approximation of the distance of the midpoint of a cell and the curve to determine whether the cell hits the curve or not. The majority of recent published algorithms perform the test in a reliable way using interval arithmetics as a tool for range analysis [8, 14, 17, 4]. To do so, $f(x)$ is evaluated with respect to the interval vector corresponding to the axes-aligned box to be tested. If the resulting interval contains zero, the box may contain a part of the object and further subdivision is performed. To reduce overestimations caused by interval arithmetic, de Figuereido and Stolfi [7] replace interval arithmetic by affine arithmetic. Voiculescu et al. [23] introduce two further methods to improve the results in the case of algebraic curves and surfaces: They show that a reformulation of the equation into Bernstein-Bezier form and/or the use of a modified affine arithmetic [13] improves the result and the performance of the subdivision algorithm.

The algorithms based on interval or affine arithmetic mentioned above perform the incidence test with the original curve or surface. This test is computationally expensive for objects of high algebraic degree or non-trivial non-algebraic description.

Furthermore, pure subdivision algorithms suffer from the high number of necessary subdivisions to reach pixel size. An adaptive subdivision algorithm loses its reliability due to the approximation implied in the linearisations representing the final results.

In this paper, a new fast and reliable adaptive subdivision algorithm for plotting implicit curves is developed.

Reconsidering the problems of existing algorithms, the new algorithm should provide

- a simplified box/object incidence test

- a simple criterion for adaptive subdivision

- a reliable representation of the results

For this purpose, the author introduces *Implicit Linear Interval Estimations* (ILIEs) for implicitly defined curves to reduce the overall costs of the enumeration algorithm and to provide a reliable enclosure of the object. ILIEs are an extension of the existing concept of parametric Linear Interval Estimations. Parametric LIEs have been successfully introduced by the author in the context of a reliable subdivision algorithm for parametric surfaces [6] and for general parametric objects [5].

Structure of the paper. After a short overview on interval and affine arithmetics in section 2.2, a definition of ILIEs for curves is given in section 2.3. In the same section, a method to compute ILIEs based on affine arithmetic is developed and a characterisation is presented. The theoretical ideas of the previous sections are applied in section 2.4 to an adaptive subdivision algorithm to enumerate and enclose an implicit curve. The paper ends with conclusions and ideas for future work.

Notations. In the following, $\mathbf{R}$ denotes the set of real numbers and $\mathbf{IR}$ the set of intervals. Furthermore, if there is no other declaration, thin small letters ($u \in \mathbf{R}$) denote real scalars, thin capital letters ($I_u \in \mathbf{IR}$) intervals, bold letters ($\mathbf{x} \in \mathbf{R}^n$) real vectors and hollow letters ($\mathbb{I}$ or $\mathbb{x} \in \mathbf{IR}^n$) interval vectors. Affine forms are marked with a hat ($\hat{u}, \hat{f}$).

2. Interval and Affine Arithmetics

Interval Arithmetic. Intervals can be used to describe fuzzy data: The wish to record errors caused by the finite precision of floating point operations gave the initial impulse to use interval arithmetic for numerical calculations. Only a short introduction to some basic ideas of interval arithmetic can be given in this paper. The books of Neumaier [12] and Alefeld et al. [1] are recommended for further reading.

Interval arithmetic operates on the set of compact intervals **IR**, where a compact interval $I = [a,b] \in \textbf{IR}$ is defined as

$$[a,b] := \{x \in \mathbf{R} \,|\, a \leq x \leq b\}$$

For $I = [a,b] \in \textbf{IR}$, $\inf(I) := a$ denotes the *infimum*, $\sup(I) := b$ the *supremum*, $rad(I) := (b-a)/2$ the *radius*, $mid(I) := (a+b)/2$ the *midpoint* and $|I| := \max\{a,b\}$ the *absolute value* of I. An interval is called *thin* if $\inf(I) = \sup(I)$.

For $I = [a,b]$, $J = [c,d]$, the basic arithmetic operations $+, -, \cdot, /$ are defined as $I + J := [a+c, b+d]$, $I - J := [a-d, b-c]$, $I \cdot J := [\min\{ac, ad, bc, bd\}, \max\{ac, ad, bc, bd\}]$. If $0 \notin J$, division is defined as $I / J := [a,b] \cdot [\frac{1}{d}, \frac{1}{c}]$.

The order relations $\sim \in \{<, \leq, \geq, >\}$ for intervals have the definition $I \sim J \Leftrightarrow x \sim y \;\; \forall x \in I, y \in J$.

If *interval arithmetics with directed roundings* is used, the result of a direct interval evaluation $\phi(I)$ of a function $\phi(x), x \in I \in \textbf{IR}$ is always an enclosure of the range $R_\phi := \{\phi(x) | x \in I\}$ of ϕ, that is overestimated in most of the cases and only optimal for some special functions.

For *interval vectors* $\mathbf{x} \in \textbf{IR}^n$ the terms infimum, supremum, midpoint, radius and absolute value, as well as the comparison and inclusion relations are used componentwise. Interval vectors are often referred to as *axes aligned boxes* to emphasise the geometric interpretation.

Reliable range analysis is an important application of interval analysis and the overestimations caused by direct interval evaluations is an often criticised drawback. Recent research in the field of reliable arithmetics tries to reduce the effect of overestimation, allowing a flexible refinement of the computation or taking more information about occurring errors into account. One of these approaches, *affine arithmetic*, has been introduced by Stolfi and de Figuereido and will be presented in the next paragraph.

Affine Arithmetic [15]. Affine arithmetic reduces the uncontrollable blow up of intervals during the evaluation of arithmetic expressions, taking dependencies of uncertainty factors of input values, approximation and rounding errors into account.

Definition [15]: *A partially unknown quantity x is represented by an* affine form

$$\hat{x} := x_0 + x_1\epsilon_1 + x_2\epsilon_2 + \cdots + x_n\epsilon_n$$

denoted in the following by the vector $(x_0, \ldots, x_n)$. The x_i are known real coefficients, the $\epsilon_i \in [-1, 1]$ are symbolic variables, standing for an independent source of error or uncertainty.

x_0 *is called the* central value *of the affine form, the x_i are the* partial deviations *and the ϵ_i the* noise symbols.

Each interval can be expressed as an affine form but an affine form can only be approximated by an interval, as it carries much more information. An interval describes only the general uncertainty of the data, whereas affine arithmetic splits this uncertainty into specific parts. Thus, a conversion from affine forms to intervals implies in most cases a loss of information.
Let $\hat{x} := x_0 + x_1\epsilon_1 + x_2\epsilon_2 + \cdots + x_n\epsilon_n$ be the affine form of the fuzzy quantity x. x lies in the interval

$$[\hat{x}] := [x_0 - \xi, x_0 + \xi]; \qquad \xi := \sum_{i=1}^{n} |x_i|$$

$[\hat{x}]$ is the smallest interval enclosing all possible values of x.

Let $X = [a, b]$ be an interval representing the value x. Then x can be represented as the affine form

$$\hat{x} = x_0 + x_k \epsilon_k$$

with $x_0 := (b + a)/2$; $x_k := (b - a)/2$.
Addition and scalar multiplication are so-called affine operations and follow simple rules applied to their evaluation with affine forms:
Let $\hat{x} = (x_0, \ldots, x_n)$ and $\hat{y} = (y_0, \ldots, y_n)$ be two affine forms with respect to the same noise symbols $\epsilon_0, \ldots, \epsilon_n$ and $\alpha \in \mathbf{R}$. Then

$$\begin{aligned}
\hat{x} \pm \hat{y} &:= (x_0 \pm y_0, \ldots, x_n \pm y_n) \\
\alpha \hat{x} &:= \alpha(x_0, \ldots, x_n) \\
\hat{x} + \alpha &:= (x_0 \pm \alpha, x_1, \ldots, x_n)
\end{aligned}$$

Non-affine operations are more difficult to determine. Stolfi and de Figueiredo [15] propose the following general strategy for the implementation: Split the operation (if possible) into an affine part and a non-affine part. Calculate the affine part as described in the previous section. For all non-affine parts calculate an affine (best) approximation (e.g. Tchebycheff approximation). The approximation error has to be multiplied with a new noise symbol and has to be added to the affine form to get the affine form of the final result.
A new noise symbol has to be introduced for round-off errors. The upper bounds of all occurring round-off errors have to be added to the partial deviation of the new symbol.

3. Implicit Linear Interval Estimations

Definition. *Let $\mathcal{F} : f(x, y) = 0$, $(x, y) \in \mathbf{R}^2$ be the implicit definition of a curve in $\mathbf{R}^2$ and*

$$L(x, y) := a\,x + b\,y + J$$

with $J \in I\!R$ and $a, b \in \mathbf{R}$.
The interval line segment inside the axes aligned box $\mathbb{I} \in I\!R^2$

$$\mathcal{L} := \{(x, y) \in \mathbb{I} \mid 0 \in L(x, y)\}$$

is called Implicit Linear Interval Estimation (ILIE)
of $\mathcal{F}$ on $\mathbb{I}$, iff for all $(x, y) \in (\mathcal{F} \cap \mathbb{I})$

$$0 \in L(x, y)$$

holds.

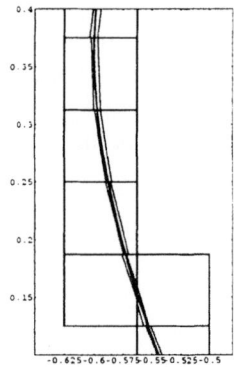

Figure 2.1. ILIEs enclosing a curve.

Thus, an ILIE of a curve on a square $\mathbb{I} \subset \mathbf{R}^2$ is a fat line segment enclosing all points of the curve inside $\mathbb{I}$. Figure 2.1 shows a set of ILIEs enclosing a curve on several boxes.

Computation of ILIEs. Affine arithmetic is used by de Figueiredo and Stolfi [7] to check whether an axes aligned box $\mathbb{I}$ hits a curve/surface $\mathcal{F} : f(x) = 0$ or not. To do so, $f(x)$ is evaluated with respect to the vector of affine forms $\hat{I}$ corresponding to $\mathbb{I}$. The resulting affine form $\hat{f}$ is converted back into an interval to determine whether zero is included in the result or not. Caused by the conversion of the result from affine form to interval, all additional information provided by the affine form is lost. An ILIE for an implicit curve can be computed with almost no additional cost using a part of this information. The following theorem gives the theoretical background for the computation of ILIEs for implicit curves in 2-space.

Theorem. *Let*

- $\mathcal{F} : f(x, y) = 0$ *be an implicit curve of $\mathbf{E}^2$*
- $\mathbb{I} = I_x \times I_y \subset \mathbf{R}^2$ *be a non-degenerated interval box*
- $\hat{x} = \hat{x}(\epsilon_x) = x_0 + x_1\,\epsilon_x$, $\hat{y} = \hat{y}(\epsilon_y) = y_0 + y_1\,\epsilon_y$ *with $\epsilon_x, \epsilon_y \in [-1, 1]$ be the corresponding affine forms to $\mathbb{I}$.*

Define

- for $\epsilon_i \in [-1,1], i = 0,\ldots,n$

$$\hat{f}(\epsilon_x, \epsilon_y, \epsilon_0, \ldots, \epsilon_n) := \hat{f}(\hat{x}, \hat{y})$$
$$= f^0 + f^x \epsilon_x + f^y \epsilon_y + \sum_{i=0}^{n} f^i \epsilon_i$$

- for $(x,y)^T \in \mathbb{I}$

$$L(x,y) := \hat{f}(\tfrac{1}{x_1}(x-x_0), \tfrac{1}{y_1}(y-y_0), [-1,1], \ldots, [-1,1])$$
$$= J + f^x \tfrac{1}{x_1} x + f^y \tfrac{1}{y_1} y$$

with $J := f^0 - \tfrac{x_0}{x_1} f^x - \tfrac{y_0}{y_1} f^y + [-\sum_{i=1}^n |f^i|, \sum_{i=1}^n |f^i|]$.

Then

$$\mathcal{L} := \{(x,y)^T \in \mathbb{I} \mid 0 \in L(x,y)\}$$

is an ILIE of $\mathcal{F}$ on $\mathbb{I}$.

Proof: All conditions and definitions of the theorem are presumed. It follows from the definition of affine forms, that the relation of (x,y) and (ϵ_x, ϵ_y)

$$\mathbb{I} \in \mathbb{R}^2 \to [-1,1]^2$$
$$(x,y)^T \mapsto (\tfrac{1}{x_1}(x-x_0), \tfrac{1}{y_1}(y-y_0))^T =: (\epsilon_x, \epsilon_y)^T$$

is a bijection iff $\mathbb{I}$ is not degenerate.
The definition of affine arithmetic guarantees that for every point $(x',y')^T \in \mathbb{I}$ there exist $\epsilon'_x, \epsilon'_y, \epsilon'_i \in [-1,1], i = 0,\ldots,n$, so that

$$f(x',y') \equiv \hat{f}(\epsilon'_x, \epsilon'_y, \epsilon'_1, \ldots, \epsilon'_n)$$

Furthermore, it follows from the inclusion property of interval arithmetics that

$$f(x',y') \in L_\epsilon(\epsilon'_x, \epsilon'_y) := \hat{f}(\epsilon'_x, \epsilon'_y, [-1,1], \ldots, [-1,1]) \quad (2.1)$$

The bijective relation of affine forms and intervals allows us to rewrite L_ϵ with respect to the coordinates (x,y)

$$L(x,y) := L_\epsilon(\tfrac{1}{x_1}(x-x_0), \tfrac{1}{y_1}(y-y_0))$$
$$= f^0 + f^x \tfrac{1}{x_1}(x-x_0) + f^y \tfrac{1}{y_1}(y-y_0))$$
$$+ [-\sum_{i=1}^n |f^i|, \sum_{i=1}^n |f^i|]$$
$$= J + f^x \tfrac{1}{x_1} x + f^y \tfrac{1}{y_1} y$$

with $J := f^0 - \frac{x_0}{x_1} f^x - \frac{y_0}{y_1} f^y + [-\sum_{i=1}^n |f^i|, \sum_{i=1}^n |f^i|]$.

Now, (2.1) can be rewritten as

$$f(x', y') \in L(x', y') \qquad (2.2)$$

$L(x, y)$ is of the form $ax + by + J$, $a, b \in \mathbf{R}$, $J \in \mathrm{IR}$ and from equation (2.2) follows for all $(x, y)^T \in \mathbb{I}$ with $f(x, y) = 0$ that $0 \in L(x, y)$.

Thus, by definition, $\mathcal{L} = \{(x, y)^T \in \mathbb{I} \mid 0 \in L(x, y)\}$ is an ILIE of $\mathcal{F}$: $f(x, y) = 0$, $(x, y)^T \in \mathbb{I}$. □

Characterisations.

1. A ILIE can be interpreted as *linearisation* of the curve inside a certain range.

2. Furthermore, the interval part of the ILIE gives reliable information of the deviation of the curve: The diameter $d := \frac{1}{\|(a,b)^T\|} diam(J)$ of $\mathcal{L}$: $ax + by + J = 0, J \in \mathrm{IR}$ can be used as an approximation of the curvature.

4. The Modified Enumeration Algorithm

The ILIEs defined in the last section are used to accelerate the subdivision process of the enumeration algorithm: In each subdivision step the new sub-boxes are tested first for incidence with the ILIE corresponding to its "mother box". Only if this test is successful the algorithm starts a new recursion step. Note that an incidence test of a box and an ILIE is very cheap compared to a curve/box incidence test with respect to the computation time due to the linear structure of the ILIE. A second advantage of ILIEs is the possibility to represent the generated results as tight piecewise linear enclosures of the curve instead of overestimating axes aligned bounding boxes (see figure 2.1).

The modified enumeration algorithm is listed in table 2.1 in pseudo code.

Experimental Results. The algorithm has been implemented in C++. The object oriented affine arithmetic package of van Iwaarden [22] has been extended. Mechanisms to distinguish between coefficients of error symbols corresponding to input intervals and coefficients of error symbols generated during the computation have been added to allow the computation of ILIEs. The interval arithmetic has been realised with the PROFIL/BIAS package [10]. Algebraic and trigonometric curves have been tested.

Test 1: Comparison of algorithms using interval or affine arithmetic for the incidence test and the **Enumeration** algorithm using ILIEs.

Fast and Reliable Plotting of Implicit Curves 23

Input: Curve object including the implicit description of the curve, an axes aligned box and the corresponding LIE.
Output: Set of ILIEs and axes-aligned boxes enclosing the curve.

    Algorithm **Enumerate** (**curve** c)
    if (0 **not** in $c.f(c.box)$)
        return;   // Stop if box does not hit the curve
    if (diameter of $c.ilie$ is small enough){
        write results;
        return;}   // Stop if termination criterion fulfilled
    subdivide $c.box$ into four sub-boxes, B_i, i=1,...,4;
    for i=1,...,4
        if (0 in $c.ilie(B_i)${
                                   // Test if B_i may hit the curve.
            c_i = **curve**(B_i);
                                   // If yes, evaluate f(B_i) with affine arithmetic
                                   // and compute new ILIEs.
        Enumerate(c_i);}   // Perform test for new parts;

Table 2.1. The modified enumeration algorithm.

Figures 2.2 and 2.3 and Table 2.2 show some results of the test for two different curves. To produce comparable results, the termination criterion for the box-based algorithms has been adapted to that for ILIEs: In the case of interval and affine arithmetic, the length of the longest edge of the box has been chosen and in the case of ILIEs, the diameter of the interval plane. Note, that in figure 2.2 ILIEs representing the result are so thin, that they appear as lines, whereas in figure 2.3 parallel lines representing the ILIEs can still be recognised. Results with respect to the same curves computed with the adapted affine arithmetic introduced by Martin and Zhang [13] can be found in [23].

Comparing computation time, number of subdivisions, and the quality of the plot, the **Enumeration** algorithm using ILIEs is better with respect to all three criteria.

Test 2: Plotting non-trivial curves. It turns out that the algorithm manages in many cases without problems singularities (see figure 2.3) : curves of high algebraic degree and complicated trigonometric curves. Figure 2.4 shows plots generated with the enumeration algorithm **Enumerate**. The precision has been chosen in a way that enclosing ILIEs appear as lines. Examples (a)-(c) can be found in previous publications of different authors and allow a direct comparison of the results. Example (d) shows an algebraic curve (a plane section of

Results for figure 2.2			Results for figure 2.3		
Method	Tested boxes	Rel. time	Method	Tested boxes	Rel. time
ILIEs	379	0.03	ILIEs	197	0.25
Interval	14485	0.27	Interval	4857	0.31
Affine	4398	1.0	Affine	1029	1.0

Table 2.2. Comparisons of necessary subdivisions and computation time of the example in figure 2.2 and 2.3.

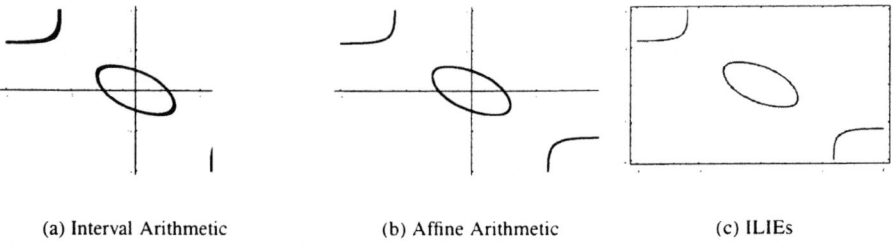

(a) Interval Arithmetic   (b) Affine Arithmetic   (c) ILIEs

Figure 2.2. Plots of $x^2 + y^2 + xy - 0.5x^2y^2 - 0.25 = 0$, $(x, y) \in [-2, 2]^2$. Precision: 0.01.

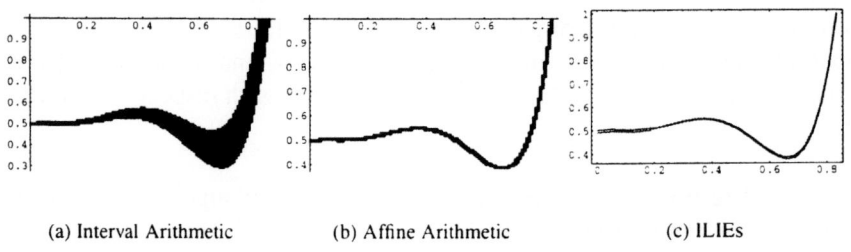

(a) Interval Arithmetic   (b) Affine Arithmetic   (c) ILIEs

Figure 2.3. Plots of $20160x^5 - 30176x^4 + 14156x^3 - 2344x^2 + 151x + 237 - 480y = 0$, $(x, y) \in [0, 1]^2$. The expression was evaluated using the Horner method. Precision: 0.01.

Fast and Reliable Plotting of Implicit Curves

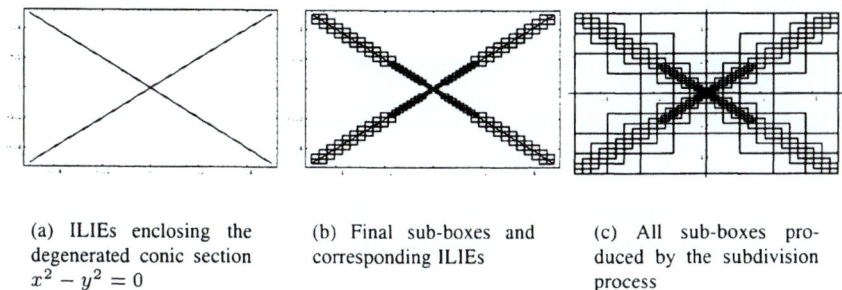

(a) ILIEs enclosing the degenerated conic section $x^2 - y^2 = 0$

(b) Final sub-boxes and corresponding ILIEs

(c) All sub-boxes produced by the subdivision process

Figure 2.4. Behaviour of the **Enumeration** algorithm at singular points.

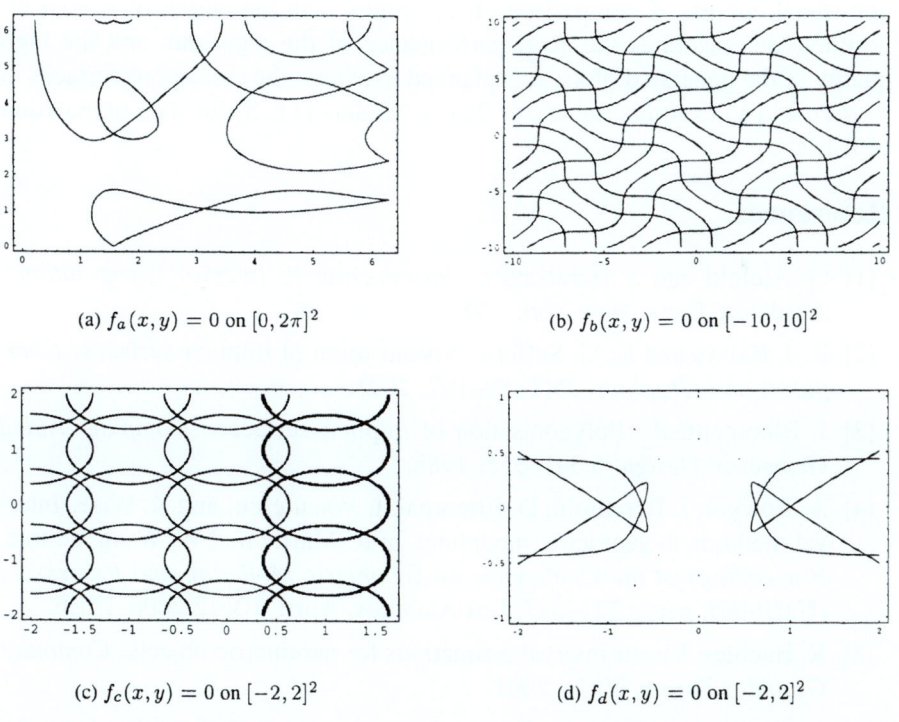

(a) $f_a(x,y) = 0$ on $[0, 2\pi]^2$

(b) $f_b(x,y) = 0$ on $[-10, 10]^2$

(c) $f_c(x,y) = 0$ on $[-2, 2]^2$

(d) $f_d(x,y) = 0$ on $[-2, 2]^2$

Figure 2.5. Plots of non-trivial implicit curves.

the Barth Decic) of degree 10 with 12 isolated singular points and 12 other singularities. The functions $f_a - f_d$ corresponding to the curves in figure 2.4 have the following form:

(a) $f_a(x,y) = \sin(x \cos y) - \cos(y \sin x)$ (see also [9]).

(b) $f_b(x,y) = \cos(\cos(\min(y + \sin x, x + \sin y))) - \cos(\sin(\max(y + \sin x, x + \sin y)))$ (see also [21]).

(c) $f_c(x,y)$ is a polynomial of degree 50. The complete description can be found in [20], page 39, figures 19 and 20, (i) to (j).

(d) $f_d(x,y) = (3+5\,t)\,(-1+x^2+y^2)^2\,(-2+t+x^2+y^2)^2 - 8\,t^4\,x^2\,y^2\,(x^2 - t^4\,y^2)\,(x^4 - 2\,x^2\,y^2 + y^4)$.

5. Conclusions and Future Work

ILIEs have been introduced to design a new enumeration and plotting algorithm for implicit curves. The algorithm has been tested with respect to several non-trivial curves. A comparison of the results with the output of similar algorithms demonstrates the good performance of the algorithm and the high quality of the generated plots. It is planned to extend the concept to surfaces in 3-space and to examine the use of Taylor Models [11, 5] for the computation of ILIEs.

References

[1] G. Alefeld and J. Herzberger. *Introduction to Interval Computations*. Academic Press, New York, 1983.

[2] R. J. Balsys and K. G. Suffern. Visualisation of implicit surfaces. *Computers and Graphics*, 25(1):89–107, 2001.

[3] J. Bloomenthal. Polygonisation of implicit surfaces. *Computer Aided Geometric Design*, 5:341–355, 1988.

[4] A. Bowyer, J. Berchtold, D. Eisenthal, I. Voiculescu, and K. Wise. Interval methods in geometric modeling. In R. Martin and W. Wang, editors, *Proceedings of the Conference on Geometric Modeling and Processing (GMP-00)*, pages 321–327, Los Alamitos, April 10–12 2000. IEEE.

[5] K. Buehler. Linear interval estimations for parametric objects. *Computer Graphics Forum*, 20(3), 2001.

[6] K. Buehler and Wilhelm Barth. A new intersection algorithm for parametric surfaces based on linear interval estimations. In *Scientific Computing, Validated Numerics, Interval Methods*. Kluwer Academic Publishers, 2001. to appear.

[7] L. H. de Figueiredo and J. Stolfi. Adaptive enumeration of implicit surfaces with affine arithmetic. *Computer Graphics Forum*, 15(5):287–296, 1996.

[8] T. Duff. Interval arithmetic and recursive subdivision for implicit functions and constructive solid geometry. *Computer Graphics*, 26(2):131–138, 1992.

[9] T. J. Hickey, Z. Qiu, and M. H. Van Emden. Interval constraint plotting for interactive visual exploration of implicitly defined relations. *Reliab. Comput.*, 6(1):81–92, 2000.

[10] W. Knueppel. Profil - programmer's runtime optimized fast interval library. Technical Report 93.4, TU Hamburg Harburg, Technische Informatik II, 1993.

[11] K. Makino and M. Berz. Efficient control of the dependency problem based on Taylor model methods. *Reliable Computing*, 5:3–12, 1999.

[12] A. Neumaier. *Interval Methods for Systems of Equations*, volume 37 of *Encyclopedia of Mathematics and its Applications*. Cambridge University Press, 1990.

[13] R.R. Martin Q. Zhang. Polynomial evaluation using affine arithmetic for curve drawing. In *Eurographics UK 2000 Conference Proceedings*, pages 49–56, Abingdon, 2000. Eurographics UK.

[14] J. M. Snyder. Interval analysis for computer graphics. *Computer Graphics*, 26(2):121–130, 1992.

[15] J. Stolfi and L. H. de Figueiredo. Self-validated numerical methods and applications, 1997. Course Notes for the 21th Brazilian Mathematics Colloquium held at IMAP, July 1997.

[16] N. Stolte and A. Kaufman. Parallel spatial enumeration of implicit surfaces using interval arithmetic for octree generation and and its direct visualization. In *Implicit Surfaces'98*, pages 81–88, 1998.

[17] K. G. Suffern and E. D. Fackerell. Interval methods in computer graphics. *Computers and Graphics*, 15(3):331–340, 1991.

[18] S. Tanaka, A. Morisaki, S. Nakata, Y. Fukuda, and H. Yamamoto. Sampling implicit surfaces based on stochastic differential equations with converging constraint. *Computers and Graphics*, 24(3):419–431, 2000.

[19] G. Taubin. An accurate algorithm for rasterizing algebraic curves. *IEEE Computer Graphics and Applications*, 14(2):14–23, 1994.

[20] G. Taubin. Distance approximations for rasterizing implicit curves. *ACM Transactions on Graphics*, 13(1):3–42, 1994.

[21] J. Tupper. Reliable two dimensional graphing methods for mathematical formulae with two free variable. In *Procceedings of the ACM Siggraph 2001*, 2001.

[22] R. van Iwaarden and J. Stolfi. Affine arithmetic sofware. 1997.

[23] I. Voiculescu, J. Berchtold, A. Bowyer, R. R. Martin, and Q. Zhang. Interval and affine arithmetic for surface location of power- and Bernstein-Form polynomials. In R. Cipolla and R. Martin, editors, *The Mathematics of Surfaces*, volume IX, pages 410–423. Springer, 2000.

Chapter 3

DATA ASSIMILATION WITH SEQUENTIAL GAUSSIAN PROCESSES *

Lehel Csato, Dan Cornford, Manfred Opper
Neural Computing Research Group, Aston University,
Birmingham B4 7ET, United Kingdom
{csatol,d.cornford,m.opper}@aston.ac.uk

Abstract We study a data assimilation problem using Gaussian processes (GPs) where the GPs act as latent variables for the observations. Inference is done using a convenient parameterisation and sequential learning for a faster algorithm. We are addressing the disadvantage of the GPs, namely the quadratic scaling of the parameters with data and eliminate the scaling by using a fixed number of parameters. The result is a sparse representation that allows us to treat problems with a large number of observations. We apply our method to the prediction of wind fields over the ocean surface from scatterometer data.

1. Introduction

Gaussian processes (GP) (Williams and Rasmussen, 1996) are widely studied in the machine learning community. They belong to the family of kernel methods (Vapnik, 1995) with the possibility of full probabilistic treatment of the problem. The applicability to real-world problems of kernel methods comes from the implicit projection operation; the projection is replaced by the scalar product between two elements in the feature space, and this bivariate function is the kernel. Using the kernels we obtain algorithms that are conceptually simple, yet produce excellent results (Schoelkopf et al., 1999).

A major drawback of the GPs is their quadratic and cubic scaling of the memory and time requirements with the data. Inference or prediction for unseen data also requires all training data (with quadratic scaling), and thus applying GPs for large datasets is not possible without approximations.

---

*This article is a revised version of "Online Learning of Wind-Field Models", published in the Proceedings of the **International Conference on Artificial Neural Networks**, Vienna, 2001.

GPs are often applied to data assimilation problems in numerical weather prediction (NWP), but ad-hoc approaches are used to fine-tune them and to overcome the enormous time and memory requirements (Daley, 1991). The aim of "data assimilation" is to combine observations and *a priori* knowledge to obtain an optimal estimate of the current state of the atmosphere (Lorenc, 1986). This current state estimate can then be used as the initial condition to produce weather forecasts for some future time. For a good overview the reader is referred to (Ide et al., 1997). We will apply sequential GPs to infer the wind-fields from scatterometer observations (Nabney et al., 2000) obtained from the ERS-2 satellite (Offiler, 1994).

The independent spatially distributed observations constitute the data; the connections between them is given by the GP prior, and the global posterior process approximates the wind-field that generates the observations. The relation of the observations to the wind-fields is one-to-many (Evans et al., 2000), making the retrieval of a wind field a complex problem with multiple solutions. A Bayesian framework for wind field retrieval where a vector GP prior is combined with local forward (wind field to scatterometer) or inverse models has been proposed in (Nabney et al., 2000). One problem with the approach outlined there is that the vector Gaussian process requires a matrix inversion which scales as N^3 where N is the size of the dataset. The backscatter is measured over $50 \times 50\ km$ cells over the ocean and the total number of observations acquired on a given orbit can be several thousand. The memory requirements for the GP has also a quadratic scale, unsuitable for large datasets.

We build a two-stage approximation. The first step uses the sequential update (Opper, 1998) followed by a second step where a geometrical argument helps us to reduce the number of parameters required for the GP (Csato and Opper, 2002), thus leading to a sparse algorithm. The time complexity of the algorithm scales linearly with the dataset size and the required memory is independent of it. The resulting algorithm is similar to the familiar Support Vectors (Vapnik, 1995), or Relevance Vectors (Tipping, 2000), with the additional possibility of probabilistic treatment and a fully scalable algorithm. This article proposes a sparse update that minimises the KL (Kullback-Leibler)–distance or relative entropy in the space of all GPs, replacing the heuristic scheme proposed earlier (Csato et al., 2001). The implications of this replacement are detailed.

The next section presents the inference of wind fields modelled as vector GPs from scatterometer observations. The Bayesian online learning and its sparse extension are detailed in Section 3.3. Section 3.3.2 proposes a measure to assess the relative weight of the different online solutions, and Section 3.4 is the discussion.

2. Processing Scatterometer Data

Scatterometers are commonly used to retrieve wind vectors over ocean surfaces. Current methods of transforming the observed values (scatterometer data, denoted as vectors s or s_i at a given spatial location) into wind fields can be split into two phases: local wind vector retrieval and ambiguity removal (Stoffelen and Anderson, 1997) where one of the local solutions is selected as the true wind vector. Ambiguity removal often uses external information, such as a NWP forecast of the expected wind field at the time of the scatterometer observations. We are seeking a method of wind field retrieval which does not require external data.

In this paper we use a mixture density network (MDN) (Bishop, 1995) to model the conditional dependence of the local wind vector $z_i = (u_i, v_i)$ on the local scatterometer observations s_i:

$$p_m(z_i|s_i, \omega) = \sum_{j=1}^{4} \beta_{ij} \phi(z_i|c_{ij}, \sigma_{ij}) \quad (3.1)$$

where ω denotes the parameters of the MDN, ϕ is a Gaussian distribution whose parameters are functions of ω and s_i. The parameters of the MDN are determined using an independent training set (Evans et al., 2000) and are considered known in this application. The MDN, which has four Gaussian component densities, captures the ambiguity of the inverse problem.

In order to have a global model from the localised wind vectors, we have to combine them. We use a zero-mean vector GP to link the local inverse models (Nabney et al., 2000):

$$q(\underline{z}) \propto \left(\prod_{i}^{N} \frac{p_m(z_i|s_i, \omega) p(s_i)}{p_G(z_i|W_{0i})} \right) p_0(\underline{z}|W_0) \quad (3.2)$$

where $\underline{z} = [z_1, \ldots, z_N]^T$ is the concatenation of the local wind field components, $W_0 = \{W_0(x_i, x_j)\}_{i,j=1,\ldots,N}$ is the prior covariance matrix for the vector $\underline{z}$ (dependent on the spatial location of the wind vectors), and p_G is p_0 marginalised at z_i, a zero-mean Gaussian with covariance W_{0i}. The choice of the kernel function $W_0(x, y)$ fully specifies our prior beliefs about the model. Notice also that for any given location we have a *two-dimensional* wind vector, and thus the output of the kernel function *is a 2×2 matrix*; details can be found in (Nabney et al., 2000). The link between two different wind field directions is made through the kernel function – the larger the kernel value, the stronger the "coupling" between the two corresponding wind fields. The prior Gaussian process is tuned carefully to represent features seen in real wind fields.

Since all quantities involved are Gaussians, we could, *in principle*, compute the resulting probabilities analytically, but this computation is *practically* in-

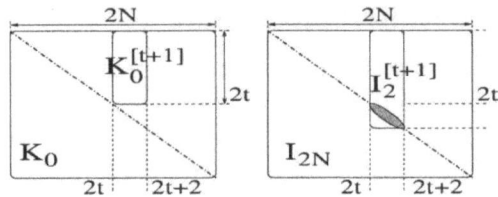

Figure 3.1. Illustration of the elements used in the update eq. (3.5).

tractable: the number of mixture elements from $q(\underline{z})$ is 4^N, extremely high even for moderate values of N. Instead, we will apply the online approximation of (Csato and Opper, 2002) to have a jointly Gaussian approximation to the posterior at all data points. However, we know that the posterior distribution of the wind field given the scatterometer observations is multi-modal, with in general two dominating and well separated modes. We might thus expect that the online implementation of the Gaussian process will track one of these posterior modes. Results show that this is indeed the case, although the order of the insertion of the local observations appears to be important.

3. Online Learning for Gaussian Processes

Gaussian processes belong to the family of Bayesian models (Bernardo and Smith, 1994). However, contrary to the finite-dimensional case, here the "model parameters" are continuous: the GP priors specify a Gaussian distribution over a function space. Due to the vector GP, the kernel function $\boldsymbol{W}_0(x, y)$ is a 2×2 matrix, specifying the pairwise cross-correlation between wind field components at different spatial positions.

Simple moments of GP posteriors (which are usually non Gaussian) have a parameterisation in terms of the training data (Opper and Winther, 1999) which resembles the popular kernel-representation (Kimeldorf and Wahba, 1971). For all spatial locations x, the mean and covariance function of the vectors $\boldsymbol{z}_x \in \mathbb{R}^2$ are represented as

$$\langle \boldsymbol{z}_x \rangle = \sum_{i=1}^{N} \boldsymbol{W}_0(x, x_i)\, \boldsymbol{\alpha}_z(i)$$
$$\mathrm{cov}(\boldsymbol{z}_x, \boldsymbol{z}_y) = \boldsymbol{W}_0(x, y) + \sum_{i,j=1}^{N} \boldsymbol{W}_0(x, x_i)\, \boldsymbol{C}_z(ij)\, \boldsymbol{W}_0(x_j, y) \quad (3.3)$$

where $\boldsymbol{\alpha}_z(1), \boldsymbol{\alpha}_z(2), \ldots, \boldsymbol{\alpha}_z(N)$ and $\{\boldsymbol{C}_z(ij)\}_{i,j=1,N}$ are parameters which will be updated sequentially by our online algorithm. Before doing so, we will (for numerical convenience) represent the vectorial process by a scalar

Data Assimilation with Sequential Gaussian Processes

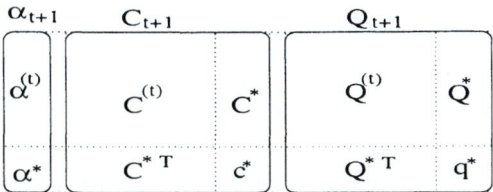

Figure 3.2. Decomposition of model parameters for the update eq. (3.8).

process with twice the number of observations, i.e. we set

$$\langle z_x \rangle = \begin{bmatrix} \langle f_{x^u} \rangle \\ \langle f_{x^v} \rangle \end{bmatrix} \quad \text{and} \quad W_0(x,y) = \begin{bmatrix} K_0(x^u, y^u) & K_0(x^u, y^v) \\ K_0(x^v, y^u) & K_0(x^v, y^v) \end{bmatrix} \quad (3.4)$$

and write (ignoring the superscripts)

$$\langle f_x \rangle = \sum_{i=1}^{2N} K_0(x, x_i) \alpha(i)$$
$$\text{cov}(f_x, f_y) = K_0(x, y) + \sum_{i,j=1}^{2N} K_0(x, x_i) C(ij) K_0(x_j, y) \quad (3.5)$$

where $\alpha = [\alpha_1, \ldots, \alpha_{2N}]^T$ and $C = \{C(ij)\}_{i,j=1,\ldots,2N}$ are rearrangements of the parameters from eq. (3.3).

The online approximation for GP learning (Csato and Opper, 2002) approximates the posterior by a Gaussian at every step. For a new observation s_{t+1}, the previous approximation to the posterior $q_t(z)$ together with a local "likelihood" factor (from eq. (3.2))

$$\frac{p_m(z_{t+1}|s_{t+1}, \omega) p(s_{t+1})}{p_G(z_{t+1}|W_{0,t+1})}$$

are combined into a new posterior using Bayes' rule. Computing its mean and covariance enable us to create an updated Gaussian approximation $q_{t+1}(z)$ at the next step. $\hat{q}(z) = q_{N+1}(z)$ is the final result of the online approximation. This process can be formulated in terms of updates for the parameters α and C which determine the mean and covariance (Csato and Opper, 2002):

$$\alpha_{t+1} = \alpha_t + v_{t+1} \frac{\partial \ln g(\langle z_{t+1} \rangle)}{\partial \langle z_{t+1} \rangle}$$
$$C_{t+1} = C_t + v_{t+1} \frac{\partial^2 \ln g(\langle z_{t+1} \rangle)}{\partial \langle z_{t+1} \rangle^2} v_{t+1}^T$$
$$v_{t+1} = C_t K_0^{[t+1]} + I_2^{[t+1]} \quad (3.6)$$

where $K_0^{[t+1]}$ and $I_2^{[t+1]}$ are shown in Fig. 3.1,

$$g(\langle z_{t+1}\rangle) = \left\langle \frac{p_m(z_{t+1}|s_{t+1},\omega)p(s_{t+1})}{p_G(z_{t+1}|W_{0,t+1})} \right\rangle_{q_t(z_{t+1})} \quad (3.7)$$

and $\langle z_{t+1}\rangle$ is a vector, implying vector and matrix first and second derivatives in eq. (3.5). The function $g(\langle z_{t+1}\rangle)$ is easy to compute analytically because it just requires the two dimensional marginal distribution of the process at the observation point s_{t+1}. Fig. 3.3 shows the results of the online algorithm applied on a sample wind field; details can be found in the figure caption.

3.1 Sparsity in Wind Fields

At each time-step, the number of nonzero parameters will be increased in the update equation. This forces us to use a further approximation which reduces the number of supporting examples in the representations eq. (3.5) to a smaller set of basis vectors. Following our approach in (Csato and Opper, 2002) we remove the last data element when a certain score (defined by the feature space geometry associated to the kernel K_0) suggests that the approximation error is small. The remaining parameters are readjusted to partly compensate for the removal as:

$$\hat{\alpha} = \alpha^{(t)} - (C^* + Q^*)(c^* + q^*)^{-1}\alpha^*$$
$$\hat{Q} = Q^{(t)} - Q^*q^{*(-1)}Q^{*T} \quad (3.8)$$
$$\hat{C} = C^{(t)} + Q^*q^{*(-1)}Q^{*T} - (C^* + Q^*)(c^* + q^*)^{-1}(C^* + Q^*)^T$$

where $Q^{-1} = \{K_0(x_i, x_j)\}_{i,j=1,\ldots,2N}$ is the inverse of the Gram matrix, the elements being shown in Fig. 3.2 (note that q^* and c^* are two-by-two matrices). The parameter updates are optimal, i.e. they are obtained by minimising the KL-divergence between the original GP and the one with the last element removed (Csato and Opper, prep). The loss when removing a data point is approximated with $\varepsilon = \|(q^* + c^*)^{-1}\alpha^*\|$. We will call it the score for the last element and the scores for other elements are obtained by permutation. Removing the data locations with low scores sequentially leaves only a small set of so-called *basis points* upon which all further prediction will depend.

The numerical experiments are promising: Fig. 3.4.a shows the resulting wind field if 85 out of the 100 spatial knots are removed from the presentation eq. (3.5). Sub-figures (b) and (c) plot the evolution of the KL-divergence and the sum-squared errors in the means between the vector GP and a trimmed GP using eq. (3.8) (dashed line) and the pruning rule proposed in (Csato et al., 2001). The horizontal axis shows the number of deleted points when the error was measured. We see that whilst the approximation of the posterior variance decays quickly, the predictive mean is fairly reliable against deleting. It is also important that the updates in eq. (3.8) give smaller errors, clearly visible if we consider the quadratic errors of the predictive means in Fig. 3.4.c.

3.2 Measuring the Relative Weight of the Approximation

An exact computation of the posterior would lead to a multi-modal distribution of wind fields at each data-point. This would correspond to a mixture of GPs as a posterior rather than to a single GP that is used in our approximation. If the individual components of the mixture are well separated, we may expect that our online algorithm will track modes with significant underlying probability mass to give a relevant prediction. However, this will depend on the actual sequence of data-points that are visited by the algorithm. To investigate the variation of our wind field prediction with the data sequence, we generated many random sequences and compared the outcomes based on a simple approximation for the relative mass of the multivariate Gaussian component.

Assuming an online solution of the marginal distribution $(\hat{\mathbf{z}}, \hat{\boldsymbol{\Sigma}})$ at a separated mode, we have the posterior at the local maximum :

$$q(\hat{\mathbf{z}}) \propto \gamma_l \, (2\pi)^{-2N/2} \, |\hat{\boldsymbol{\Sigma}}|^{-1/2} \qquad (3.9)$$

with $q(\hat{\mathbf{z}})$ from eq. (3.2), γ_l *the weight of the component* of the mixture to which our online algorithm had converged, and we assume the local curvature is also well approximated by $\hat{\boldsymbol{\Sigma}}$.

Having two different online solutions $(\hat{\mathbf{z}}_1, \hat{\boldsymbol{\Sigma}}_1)$ and $(\hat{\mathbf{z}}_2, \hat{\boldsymbol{\Sigma}}_2)$, we find from eq. (3.9) that the proportion of the two weights is given by

$$\frac{\gamma_1}{\gamma_2} = \frac{q(\hat{\mathbf{z}}_1)|\hat{\boldsymbol{\Sigma}}_1|^{1/2}}{q(\hat{\mathbf{z}}_2)|\hat{\boldsymbol{\Sigma}}_2|^{1/2}}. \qquad (3.10)$$

This helps us to estimate, up to an additive constant, the "relative weight" of the wind field solutions, helping us to assess the quality of the approximation we arrived at. Results, using multiple runs on a wind field data confirm this expectation. The correct solution (Fig. 3.3.b) has large value and high frequency if doing multiple runs. Running the algorithm using different inclusion orders leads in most cases to the two symmetric solutions identified by MCMC methods. These two solutions had similar relative weights. The wind fields at the local minimum (eg. from Fig. 3.4.d) were insignificant, but we envisage a larger-scale comparison of the results.

4. Discussion

In the wind field example the online and sparse approximation allows us to tackle much larger wind fields than previously possible. This suggests that we will be able to retrieve wind fields using only scatterometer observations, by utilising all available information in the signal.

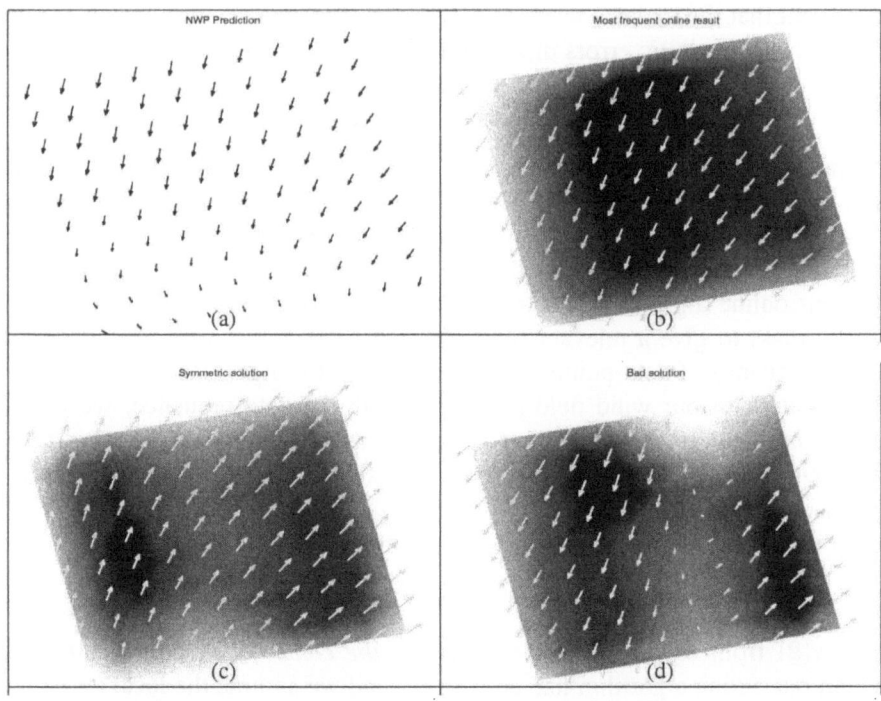

Figure 3.3. The NWP wind field estimation. (a) The most frequent. (b) The second most frequent. (c) Online solution. (d) A bad solution. The assessment of a good/bad solution is based on the value of the relative weight from Section 3.3.2. The gray-scale background indicates the model confidence (Bayesian error-bars) in the prediction, darker shade meaning more confidence.

At present we obtain different solutions for different orderings of the data. Future work might seek to build an adaptive classifier that works on the family of online solutions and utilise the relative weights. However, a more desirable method would be to extend our online approach to mixtures of GPs in order to incorporate the multi-modality of the posterior process in a principled way.

Proceeding with the removal of the basis points, it would be desirable to have an improved update for the vector GP that leads to a better estimation of the posterior kernel (and thus of the Bayesian error-bars). The KL-based updates constitute a first step in this direction. A more precise estimation of the GP parameters would be possible if we extend the online algorithm to iterate multiple times across the data, similar to the expectation-propagation algorithm (Minka, 2000).

The resulting learning algorithm is sensitive to the symmetries in the physical forward model, and a partial solution is to use a non-zero mean prior GP as is often done in NWP data assimilation (Lorenc, 1986).

Data Assimilation with Sequential Gaussian Processes

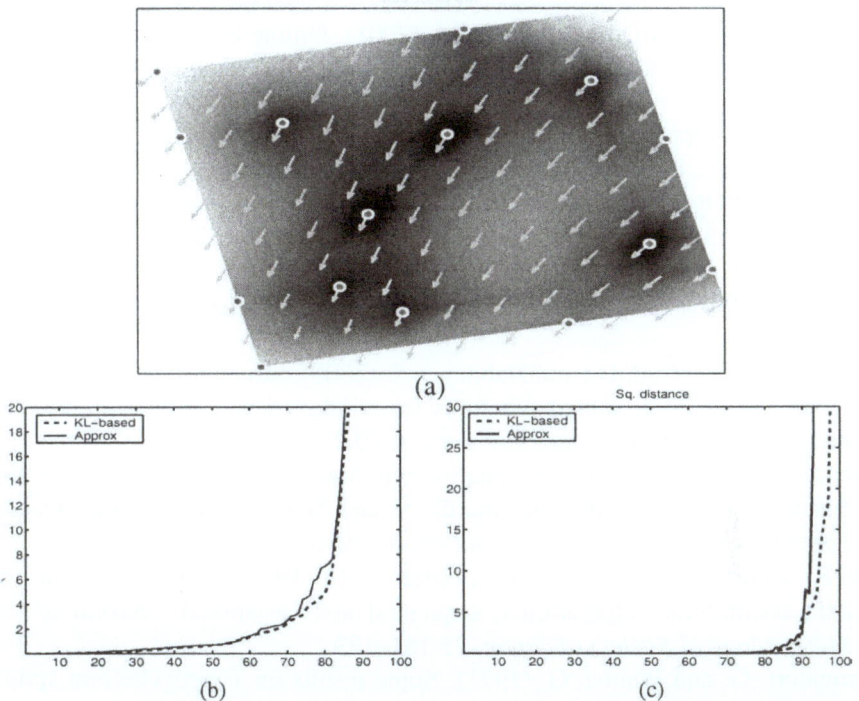

Figure 3.4. (a) The predicted wind fields when 85% of the modes has been removed (from Fig. 3.3). The prediction is based only on basis vectors (circles). The model confidence is higher at these regions. (b,c) The evolution of the KL-divergence (b) and the squared error (c) during the removal of basis vectors (the number of removed basis vectors is the X-axis). The continuous line marks the previous results based on the approximation used in (Csato et al., 2001) while the dotted line gives the results for the KL-based technique introduced in this paper.

An exciting research direction would be the inclusion of dynamics and hidden states that propagate in time. This would lead to learning methods similar to the different flavours of the extended Kalman filtering techniques: kernel-based (Roweis and Ghahramani, 2001), the recent esemble Kalman filter (Evensen, 2001), or generalised Kalman filter (Wolf, 1999).

Acknowledgement
This work was supported by EPSRC grant no. GR/M81608.

References
Bernardo, J. M. and Smith, A. F. (1994). *Bayesian Theory*. John Wiley & Sons.
Bishop, C. M. (1995). *Neural Networks for Pattern Recognition*. Oxford University Press, New York, N.Y.

Csato, L., Cornford, D., and Opper, M. (2001). Online learning of wind-field models. In *International Conference on Artificial Neural Networks*, pages 300–307.

Csato, L. and Opper, M. (2002). Sparse on-line Gaussian Processes. *Neural Computation*, 14(3):641–669.

Csato, L. and Opper, M. (prep). Greedy sparse approximation to Gaussian Processes by relative entropy projection. Technical report, Neural Computing Research Group.

Daley, R. (1991). *Atmospheric Data Analysis*. Cambridge University Press, Cambridge.

Evans, D. J., Cornford, D., and Nabney, I. T. (2000). Structured neural network modelling of multi-valued functions for wind retrieval from scatterometer measurements. *Neurocomputing Letters*, 30:23–30.

Evensen, G. (2001). Sequential data assimilation for nonlinear dynamics: the ensemble Kalmam Filter. In Pinardi, N. and Woods, J. D., editors, *Ocean Forecasting: Conceptual basis and applications*. Springer-Verlag.

Ide, K., Courtier, P., Ghil, M., and Lorenc, A. C. (1997). Unified notation for data assimilation: Operational, sequential and variational. *Journal of the Meteorological Society of Japan*, 75:181–189.

Kimeldorf, G. and Wahba, G. (1971). Some results on Tchebycheffian spline functions. *J. Math. Anal. Applic.*, 33:82–95.

Lorenc, A. C. (1986). Analysis methods for numerical weather prediction. *Quarterly Journal of the Royal Meteorological Society*, 112:1177–1194.

Minka, T. P. (2000). *Expectation Propagation for Approximate Bayesian Inference*. PhD thesis, Dep. of Electrical Eng. and Comp. Sci.; MIT.

Nabney, I. T., Cornford, D., and Williams, C. K. I. (2000). Bayesian inference for wind field retrieval. *Neurocomputing Letters*, 30:3–11.

Offiler, D. (1994). The calibration of ERS-1 satellite scatterometer winds. *Journal of Atmospheric and Oceanic Technology*, 11:1002–1017.

Opper, M. (1998). A Bayesian approach to online learning. In *On-Line Learning in Neural Networks*, pages 363–378. Cambridge Univ. Press.

Opper, M. and Winther, O. (1999). Gaussian processes and SVM: Mean field results and leave-one-out estimator. In Smola, A., Bartlett, P., Schoelkopf, B., and Schuurmans, C., editors, *Advances in Large Margin Classifiers*, pages 43–65. The MIT Press, Cambridge, MA.

Roweis, S. and Ghahramani, Z. (2001). An EM algorithm for identification of nonlinear dynamical systems. In Haykin, S., editor, *Kalman Filtering and Neural Networks*. Wiley.

Schoelkopf, B., Burges, C. J., and Smola, A. J., editors (1999). *Advances in kernel methods (Support Vector Learning)*. The MIT Press.

Stoffelen, A. and Anderson, D. (1997). Ambiguity removal and assimiliation of scatterometer data. *Quarterly Journal of the Royal Meteorological Society*, 123:491–518.

Tipping, M. (2000). The Relevance Vector Machine. In Solla, S. A., Leen, T. K., and Mueller, K.-R., editors, *NIPS*, volume 12, pages 652–658. The MIT Press.

Vapnik, V. N. (1995). *The Nature of Statistical Learning Theory*. Springer-Verlag, New York, NY.

Williams, C. K. I. and Rasmussen, C. E. (1996). Gaussian processes for regression. In Touretzky, D. S., Mozer, M. C., and Hasselmo, M. E., editors, *NIPS*, volume 8. The MIT Press.

Wolf, D. R. (1999). A Bayesian reflection on surfaces. *Entropy*, 1(4):69–98.

Chapter 4

CONFORMAL GEOMETRY, EUCLIDEAN SPACE AND GEOMETRIC ALGEBRA

Chris Doran, Anthony Lasenby, Joan Lasenby
C.Doran@mrao.cam.ac.uk
anthony@mrao.cam.ac.uk
jl@eng.cam.ac.uk
Cambridge University, Cambridge, United Kingdom

Abstract   Projective geometry provides the preferred framework for most implementations of Euclidean space in graphics applications. Translations and rotations are linear transformations in projective geometry, which helps when it comes to programming complicated geometrical operations. But there is a fundamental weakness in this approach — the Euclidean distance between points is not handled in a straightforward manner. Here we discuss a solution to this problem, based on conformal geometry. The language of geometric algebra is best suited to exploiting this geometry, as it handles the interior and exterior products in a single, unified framework. A number of applications are discussed, including a compact formula for reflecting a line off a general spherical surface.

Keywords: Geometric algebra, Clifford algebra, conformal geometry, projective geometry, homogeneous coordinates, sphere geometry, stereographic projection

1.   Introduction

In computer graphics programming the standard framework for modeling points in space is via a projective representation. So, for handling problems in three-dimensional geometry, points in Euclidean space x are represented projectively as rays or vectors in a four-dimensional space,

$$X = x + e_4. \qquad (4.1)$$

The additional vector e_4 is orthogonal to x, $e_4 \cdot x = 0$, and is normalised to 1, $(e_4)^2 = 1$. From the definition of X it is apparent that e_4 is the projective representation of the origin in Euclidean space. The projective representation is *homogeneous*, so both X and λX represent the same point. Projective space is also not a linear space, as the zero vector is excluded. Given a vector A in projective space, the Euclidean point a is then recovered from

$$a = \frac{A - A \cdot e_4 \, e_4}{A \cdot e_4}. \tag{4.2}$$

The components of A define a set of homogeneous coordinates for the position a.

The advantage of the projective framework is that the group of Euclidean transformations (translations, reflections and rotations) is represented by a set of linear transformations of projective vectors. For example, the Euclidean translation $x \mapsto x + a$ is described by the matrix transformation

$$\begin{pmatrix} 1 & 0 & 0 & a_1 \\ 0 & 1 & 0 & a_2 \\ 0 & 0 & 1 & a_3 \\ 0 & 0 & 0 & 1 \end{pmatrix} \begin{pmatrix} x_1 \\ x_2 \\ x_3 \\ 1 \end{pmatrix} = \begin{pmatrix} x_1 + a_1 \\ x_2 + a_2 \\ x_3 + a_3 \\ 1 \end{pmatrix}. \tag{4.3}$$

This linearisation of a translation ensures that compounding a sequence of translations and rotations is a straightforward exercise in projective geometry. All one requires for applications is a fast engine for multiplying together 4×4 matrices.

The main operation in projective geometry is the *exterior product*, originally introduced by Grassmann in the nineteenth century [1, 2]. This product is denoted with the wedge symbol $\wedge$. The outer product of vectors is associative and totally antisymmetric. So, for example, the outer product of two vectors A and B is the object $A \wedge B$, which is a rank-2 antisymmetric tensor or *bivector*. The components of $A \wedge B$ are

$$(A \wedge B)_{ij} = A_i B_j - A_j B_i. \tag{4.4}$$

The exterior product defines the *join* operation in projective geometry, so the outer product of two points defines the line between them, and the outer product of three points defines a plane. In this scheme a line in three dimensions is then described by the 6 components of a bivector. These are the Pluecker coordinates of a line. The associativity and antisymmetry of the outer product ensure that

$$(A \wedge B) \wedge (A \wedge B) = A \wedge B \wedge A \wedge B = 0, \tag{4.5}$$

which imposes a single quadratic condition on the coordinates of a line. This is the Pluecker condition.

The ability to handle straight lines and planes in a systematic manner is essential to practically all graphics applications, which explains the popularity of the projective framework. But there is one crucial concept which is missing.

This is the Euclidean *distance* between points. Distance is a fundamental concept in the Euclidean world which we inhabit and are usually interested in modeling. But distance cannot be handled elegantly in the projective framework, as projective geometry is non-metrical. Any form of distance measure must be introduced via some additional structure. One way to proceed is to return to the Euclidean points and calculate the distance between these directly. Mathematically this operation is distinct from all others performed in projective geometry, as it does not involve the exterior product (or duality). Alternatively, one can follow the route of classical planar projective geometry and define the additional metric structure through the introduction of the *absolute conic* [3]. But this structure requires that all coordinates are complexified, which is hardly suitable for real graphics applications. In addition, the generalisation of the absolute conic to three-dimensional geometry is awkward.

There is little new in these observations. Grassmann himself was dissatisfied with an algebra based on the exterior product alone, and sought an algebra of points where distances are encoded in a natural manner. The solution is provided by the *conformal model* of Euclidean geometry, originally introduced by Mobius in his study of the geometry of spheres. The essential new feature of this space is that it has mixed signature, so the inner product is not positive definite. In the nineteenth century, when these developments were initiated, mixed signature spaces were a highly original and somewhat abstract concept. Today, however, physicists and mathematicians routinely study such spaces in the guise of special relativity, and there are no formal difficulties when computing with vectors in these spaces. As a route to understanding the conformal representation of points in Euclidean geometry we start with a description of the *stereographic projection*. This map provides a means of representing points as null vectors in a space of two dimensions higher than the Euclidean base space. This is the conformal representation. The inner product of points in this space recovers the Euclidean distance, providing precisely the framework we desire. The outer product extends the range of geometric primitives from projective geometry to include circles and spheres, which has many applications.

The conformal model of Euclidean geometry makes heavy use of both the interior and exterior products. As such, it is best developed in the language of *geometric algebra* — a universal language for geometry based on the mathematics of *Clifford algebra* [4, 5, 6]. This is described in section 4.3. The power of the geometric algebra development becomes apparent when we discuss the group of conformal transformations, which include Euclidean transformations as a subgroup. As in the projective case, all Euclidean transformations are linear transformations in the conformal framework. Furthermore, these transformations are all *orthogonal*, and can be built up from primitive reflections.

The join operation in conformal space generalises the join of projective geometry. Three points now define a line, which is the circle connecting the points. If this circle passes through the point at infinity it is a straight line. Similarly, four points define a sphere, which reduces to a plane when its radius is infinite. These new geometric primitives provide a range of intersection and reflection operations which dramatically extend the available constructions in projective geometry. For example, reflecting a line in a sphere is encoded in a simple expression involving a pair of elements in geometric algebra. Working in this manner one can write computer code for complicated geometrical operations which is robust, elegant and highly compact. This has many potential applications for the graphics industry.

2. Stereographic projection and conformal space

The stereographic projection provides a straightforward route to the principle construction of the conformal model — the representation of a point as a null vector in conformal space. The stereographic projection maps points in the Euclidean space $\mathbb{R}^n$ to points on the unit sphere S^n, as illustrated in figure 4.1. Suppose that the initial point is given by $x \in \mathbb{R}^n$, and we write

$$x = r\hat{x}, \tag{4.6}$$

where r is the magnitude of the vector x, $r^2 = x^2$. The corresponding point on the sphere is

$$S(x) = \cos\theta\,\hat{x} - \sin\theta\,e, \tag{4.7}$$

where e is the unit vector perpendicular to the plane defining the south pole of the sphere S^n. The angle θ, $-\pi/2 \leq \theta \leq \pi/2$, is related to the distance r by

$$r = \frac{\cos\theta}{1 + \sin\theta}, \tag{4.8}$$

which inverts to give

$$\cos\theta = \frac{2r}{1 + r^2}, \quad \sin\theta = \frac{1 - r^2}{1 + r^2}. \tag{4.9}$$

The stereographic projection S maps $\mathbb{R}^n$ into $S^n - P$, where P is the south pole of S^n. We complete the space S^n by letting the south pole represent the point at infinity. We therefore expect that, under Euclidean transformations of $\mathbb{R}^n$, the point at infinity should remain invariant.

We now have a representation of points in $\mathbb{R}^n$ with unit vectors in the space $\mathbb{R}^{n+1}$. But the constraint that the vector has unit magnitude means that this representation is not homogeneous. A homogeneous representation of geometric objects is critical to the power of projective geometry, as it enables us to write the equation of the line through A and B as

$$A \wedge B \wedge X = 0. \tag{4.10}$$

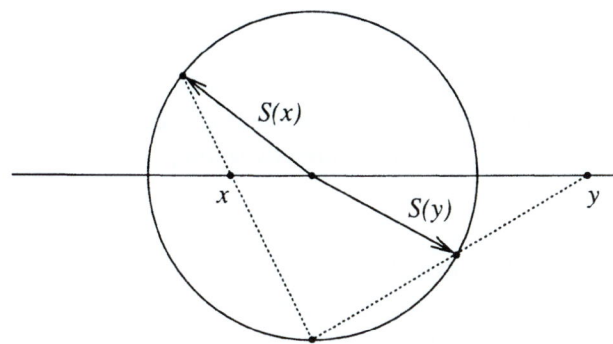

Figure 4.1. A stereographic projection. The space $\mathbb{R}^n$ is mapped to the unit sphere S^n. Given a point in $\mathbb{R}^n$ we form the line through this point and the south pole of the sphere. The point where this line intersects the sphere defines the image of the projection.

Clearly, if X satisfies this equation, then so to does λX. To achieve a homogeneous representation we introduce a further vector, $\bar{e}$, which has *negative signature*,

$$\bar{e}^2 = -1. \tag{4.11}$$

We also assume that $\bar{e}$ is orthogonal to x and e. We can now replace the unit vector $S(x)$ with the vector X, where

$$X = S(x) + \bar{e} = \frac{2x}{1+x^2} - \frac{1-x^2}{1+x^2}e + \bar{e}. \tag{4.12}$$

The vector X satisfies

$$X \cdot X = 0, \tag{4.13}$$

so is *null*. This equation is homogeneous, so we can now move to a homogeneous encoding of points and let X and λX represent the *same* point in $\mathbb{R}^n$. Multiplying the vector in equation (4.12) by $(1+x^2)$ we establish the conformal representation

$$F(x) = X = 2x - (1-x^2)e + (1+x^2)\bar{e}. \tag{4.14}$$

The vectors e and $\bar{e}$ extend the Euclidean space $\mathbb{R}^n$ to a space with two extra dimensions and signature $(n+1,1)$. It is generally more convenient to work with a null basis for the extra dimensions, so we define

$$n = e + \bar{e} \qquad \bar{n} = e - \bar{e}. \tag{4.15}$$

These vectors satisfy

$$n^2 = \bar{n}^2 = 0, \qquad n \cdot \bar{n} = 2. \tag{4.16}$$

The vector X is now

$$F(x) = X = 2x + x^2 n - \bar{n}, \qquad (4.17)$$

which defines our standard representation of points as vectors in conformal space. Given a general, unnormalised null vector in conformal space, the standard form of equation (4.17) is recovered by setting

$$X \mapsto -2\frac{X}{X \cdot n}. \qquad (4.18)$$

This map makes it clear that the null vector n now represents the point at infinity. In general we will not assume that our points are normalised, so the components of X are homogeneous coordinates for the point x. The step of normalising the representation is only performed if the actual Euclidean point is required.

Given two null vectors X and Y, in the form of equation (4.17), their inner product is

$$\begin{aligned} X \cdot Y &= (x^2 n + 2x - \bar{n}) \cdot (y^2 n + 2y - \bar{n}) \\ &= -2x^2 - 2y^2 + 4x \cdot y \\ &= -2(x-y)^2. \end{aligned} \qquad (4.19)$$

This is the key result which justifies the conformal model approach to Euclidean geometry. The inner product in conformal space encodes the *distance* between points in Euclidean space. This is why points are represented with null vectors — the distance between a point and itself is zero. Since equation (4.19) was appropriate for normalised points, the general expression relating Euclidean distance to the conformal inner product is

$$|x-y|^2 = -2\frac{X \cdot Y}{X \cdot n \, Y \cdot n}. \qquad (4.20)$$

This is manifestly homogeneous in X and Y. This formula returns the dimensionless distance. To introduce dimensions one requires a fundamental length scale λ, so that x is the dimensionless representation of the position vector λx. Appropriate factors of λ can then be inserted when required.

An orthogonal transformation in conformal space will ensure that a null vector remains null. Such a transformation therefore maps points to points in Euclidean space. This defines the full conformal group of Euclidean space, which is isomorphic to the group $SO(n+1, 1)$. Conformal transformations leave angles invariant, but can map straight lines into circles. The Euclidean group is the subgroup of the conformal group which leaves the Euclidean distance invariant. These transformations include translations and rotations, which are therefore linear, orthogonal transformations in conformal space. The key to developing simple representations of conformal transformations is geometric algebra, which we now describe.

3. Geometric algebra

The language of geometric algebra can be thought of as Clifford algebra with added geometric content. The details are described in greater detail elsewhere [4, 5, 6], and here we just provide a brief introduction. A geometric algebra is constructed on a vector space with a given inner product. The *geometric* product of two vectors a and b is defined to be associative and distributive over addition, with the additional rule that the (geometric) square of any vector is a scalar,

$$aa = a^2 \in \mathbb{R}. \tag{4.21}$$

If we write

$$ab + ba = (a+b)^2 - a^2 - b^2, \tag{4.22}$$

we see that the symmetric part of the geometric product of any two vectors is also a scalar. This defines the inner product, and we write

$$a \cdot b = \tfrac{1}{2}(ab + ba). \tag{4.23}$$

The geometric product of two vectors can now be written

$$ab = a \cdot b + a \wedge b, \tag{4.24}$$

where the exterior product is the antisymmetric combination

$$a \wedge b = \tfrac{1}{2}(ab - ba). \tag{4.25}$$

Under the geometric product, orthogonal vectors anticommute and parallel vectors commute. The product therefore encodes the basic geometric relationships between vectors. The totally antisymmetrised sum of geometric products of vectors defines the exterior product in the algebra.

Once one knows how to multiply together vectors it is a straightforward exercise to construct the entire geometric algebra of a vector space. General elements of this algebra are called *multivectors*, and they too can be multiplied via the geometric product. The two algebras which concern us in this paper are the algebras of conformal vectors for the Euclidean plane and three-dimensional space [7, 8, 9]. For the Euclidean plane, let $\{e_1, e_2\}$ denote an orthonormal basis set, and write $e_0 = \bar{e}$, $e_3 = e$. The full basis set is then $\{e_i\}, i = 0 \ldots 3$. These generators satisfy

$$e_i e_j + e_j e_i = 2\eta_{ij}, \qquad i,j = 0 \ldots 3, \tag{4.26}$$

where

$$\eta_{ij} = \text{diag}(-1, 1, 1, 1). \tag{4.27}$$

The algebra generated by these vectors consists of the set

1	$\{e_i\}$	$\{e_i \wedge e_j\}$	$\{Ie_i\}$	I
1 scalar	4 vectors	6 bivectors	4 trivectors	1 pseudoscalar.

Scalars are assigned grade zero, vectors grade one, and so on. The highest grade term in the algebra is the pseudoscalar I,

$$I = e_0 e_1 e_2 e_3 = e_1 e_2 \bar{e} e. \tag{4.28}$$

This satisfies

$$I^2 = e_1 e_2 \bar{e} e e_1 e_2 \bar{e} e = -e_1 e_2 \bar{e} e_1 e_2 \bar{e} = e_1 e_2 e_1 e_2 = -e_1 e_1 = -1. \tag{4.29}$$

The steps in this calculation involve counting the number of sign changes as a vector is anticommuted past orthogonal vectors. This is essentially how all products in geometric algebra are calculated, and it is easily incorporated into any programming language. In even dimensions, such as the case here, the pseudoscalar anticommutes with vectors,

$$Ia = -aI, \quad \text{even dimensions.} \tag{4.30}$$

For the case of the conformal algebra of Euclidean three-dimensional space $\mathbb{R}^3$, we can define a basis as $\{e_i\}, i = 0 \ldots 4$, with $e_0 = \bar{e}$, $e_4 = e$, and $e_i, i = 1 \ldots 3$, a basis set for three-dimensional space. The algebra generated by these vectors has 32 terms, and is spanned by

	1	$\{e_i\}$	$\{e_i \wedge e_j\}$	$\{e_i \wedge e_j \wedge e_k\}$	$\{Ie_i\}$	I
grade	0	1	2	3	4	5
dimension	1	5	10	10	5	1.

The dimensions of each subspace are given by the binomial coefficients. Each subspace has a simple geometric interpretation in conformal geometry. The pseudoscalar for five-dimensional space is again denoted by I, and this time is defined by

$$I = e_0 e_1 e_2 e_3 e_4 = e_1 e_2 e_3 e \bar{e}. \tag{4.31}$$

In five-dimensional space the pseudoscalar commutes with all elements in the algebra. The $(4, 1)$ signature of the space implies that the pseudoscalar satisfies

$$I^2 = -1. \tag{4.32}$$

So, algebraically, I has the properties of a unit imaginary, though in geometric algebra it plays a definite geometric role. In a general geometric algebra, multiplication by the pseudoscalar performs the duality transformation familiar in projective geometry.

4. Reflections and rotations in geometric algebra

Suppose that the vectors a and m represent two lines from a common origin in Euclidean space, and we wish to reflect the vector a in the hyperplane perpendicular to m. If we assume that m is normalised to $m^2 = 1$, the result of this reflection is

$$a \mapsto a' = a - 2(a \cdot m)m. \tag{4.33}$$

Conformal Geometry, Euclidean Space and Geometric Algebra

This is the standard expression one would write down without access to the geometric product. But with geometric algebra at our disposal we can expand a' into

$$a' = a - (am + ma)m = a - am^2 - mam = -mam. \quad (4.34)$$

The advantage of this representation of a reflection is that we can easily chain together reflections into a series of geometric products. So two reflections, one in m followed by one in l, produce the transformation

$$a \mapsto lmaml. \quad (4.35)$$

But two reflections generate a rotation, so a rotation in geometric algebra can be written in the simple form

$$a \mapsto Ra\tilde{R}, \quad (4.36)$$

where

$$R = lm, \quad \tilde{R} = ml. \quad (4.37)$$

The tilde denotes the operation of *reversion*, which reverses the order of vectors in any series of geometric products. Given a general multivector A we can decompose it into terms of a unique grade by writing

$$A = A_0 + A_1 + A_2 + \cdots, \quad (4.38)$$

where A_r denotes the grade-r part of A. The effect of the reverse on A is then

$$\tilde{A} = A_0 + A_1 - A_2 - A_3 + A_4 + \cdots. \quad (4.39)$$

The geometric product of an even number of positive norm unit vectors is called a rotor. These satisfy $R\tilde{R} = 1$ and generate rotations. A rotor can be written as

$$R = \pm \exp(B/2), \quad (4.40)$$

where B is a bivector. The space of bivectors is therefore the space of generators of rotations. These define a *Lie algebra*, with the rotors themselves defining a *Lie group*. The action of this group on vectors is defined by equation (4.36), so both R and $-R$ define the same rotation.

5. Euclidean and conformal transformations

Transformations of conformal vectors which leave the product of equation (4.20) invariant correspond to the group of Euclidean transformations in $\mathbb{R}^n$. To simplify the notation, we let $\mathcal{G}(p,q)$ denote the geometric algebra of a vector space with signature (p,q). The Euclidean spaces of interest to us therefore have the algebras $\mathcal{G}(2,0)$ and $\mathcal{G}(3,0)$ associated with them. The corresponding conformal algebras are $\mathcal{G}(3,1)$ and $\mathcal{G}(4,1)$ respectively. Each Euclidean algebra is a subalgebra of the associated conformal algebra.

The operations which mainly interest us here are translations and rotations. The fact that translations can be treated as orthogonal transformations is a novel feature of conformal geometry. This is possible because the underlying orthogonal group is non-compact, and so contains null generators. To see how this allows us to describe a translation, consider the rotor

$$R = T_a = e^{na/2}, \tag{4.41}$$

where a is a vector in the Euclidean space, so that $a \cdot n = 0$. The bivector generator satisfies

$$(na)^2 = -anna = 0, \tag{4.42}$$

so is null. The Taylor series for T_a therefore terminates after two terms, leaving

$$T_a = 1 + \frac{na}{2}. \tag{4.43}$$

The rotor T_a transforms the null vectors n and $\bar{n}$ into

$$T_a n \tilde{T}_a = n + \tfrac{1}{2}nan + \tfrac{1}{2}nan + \tfrac{1}{4}nanan = n, \tag{4.44}$$

and

$$T_a \bar{n} \tilde{T}_a = \bar{n} - 2a - a^2 n. \tag{4.45}$$

As expected, the point at infinity remains at infinity, whereas the origin is transformed to the vector a. Acting on a vector $x \in \mathcal{G}(n,0)$ we similarly obtain

$$T_a x \tilde{T}_a = x + n(a \cdot x). \tag{4.46}$$

Combining these results we find that

$$\begin{aligned} T_a F(x) \tilde{T}_a &= x^2 n + 2(x + a \cdot x\, n) - (\bar{n} - 2a - a^2 n) \\ &= (x+a)^2 n + 2(x+a) - \bar{n} \\ &= F(x+a), \end{aligned} \tag{4.47}$$

which performs the conformal version of the translation $x \mapsto x + a$. Translations are handled as rotations in conformal space, and the rotor group provides a double-cover representation of a translation. The identity

$$\tilde{T}_a = T_{-a}, \tag{4.48}$$

ensures that the inverse transformation in conformal space corresponds to a translation in the opposite direction, as required.

Similarly, as discussed above, a rotation in the origin in $\mathbb{R}^n$ is performed by $x \mapsto x' = Rx\tilde{R}$, where R is a rotor in $\mathcal{G}(n,0)$. The conformal vector representing the transformed point is

$$\begin{aligned} F(x') &= x'^2 n + 2Rx\tilde{R} - \bar{n} \\ &= R(x^2 n + 2x - \bar{n})\tilde{R} \\ &= RF(x)\tilde{R}. \end{aligned} \quad (4.49)$$

This holds because R is an even element in $\mathcal{G}(n,0)$, so must commute with both n and $\bar{n}$. Rotations about the origin therefore take the same form in either space. But suppose instead that we wish to rotate about the point $a \in \mathbb{R}^n$. This can be achieved by translating a to the origin, rotating, and then translating forward again. In terms of $X = F(x)$ the result is

$$X \mapsto T_a R T_{-a} X \tilde{T}_{-a} \tilde{R} \tilde{T}_a = R' X \tilde{R}. \quad (4.50)$$

The rotation is now controlled by the rotor R', where

$$R' = T_a R \tilde{T}_a = \left(1 + \frac{na}{2}\right) R \left(1 + \frac{an}{2}\right). \quad (4.51)$$

The conformal model frees us up from treating the origin as a special point. Rotations about any point are handled with rotors in the same manner. Similar comments apply to reflections, though we will see shortly that the range of possible reflections is enhanced in the conformal model. The Euclidean group is a subgroup of the full conformal group, which consists of transformations which preserve angles alone. This group is described in greater detail elsewhere [5, 9]. The essential property of a Euclidean transformation is that the point at infinity is invariant, so all Euclidean transformations map n to itself.

6. Geometric primitives in conformal space

In the conformal model, points in Euclidean space are represented homogeneously by null vectors in conformal space. As in projective geometry, a multivector $L \in \mathcal{G}(n+1,1)$ encodes a geometric object in $\mathbb{R}^n$ via the equations

$$L \wedge X = 0, \quad X^2 = 0. \quad (4.52)$$

One result we can exploit is that $X^2 = 0$ is unchanged if $X \mapsto RX\tilde{R}$, where R is a rotor in $\mathcal{G}(n+1,1)$. So, if a geometric object is specified by L via equation (4.52), it follows that

$$R(L \wedge X)\tilde{R} = (RL\tilde{R}) \wedge (RX\tilde{R}) = 0. \quad (4.53)$$

We can therefore transform the object L with a general element of the full conformal group to obtain a new object. As well as translations and rotations, the conformal group includes dilations and inversions, which map straight lines into circles. The range of geometric primitives is therefore extended from the projective case, which only deals with straight lines.

The first case to consider is a pair of null vectors A and B. Their inner product describes the Euclidean distance between points, and their outer product defines the bivector
$$G = A \wedge B. \tag{4.54}$$
The bivector G has magnitude
$$G^2 = (AB - A \cdot B)(-BA + A \cdot B) = (A \cdot B)^2, \tag{4.55}$$
which shows that G is *timelike*, in the terminology of special relativity. It follows that G contains a pair of null vectors. If we look for solutions to the equation
$$G \wedge X = 0, \quad X^2 = 0, \tag{4.56}$$
the only solutions are the two null vectors contained in G. These are precisely A and B, so the bivector encodes the two points directly. In the conformal model, no information is lost in forming the exterior product of two null vectors. Frequently, bivectors are obtained as the result of intersection algorithms, such as the intersection of two circles in a plane. The sign of the square of the resulting bivector, B^2, defines the number of intersection points of the circles. If $B^2 > 0$ then B defines two points, if $B^2 = 0$ then B defines a single point, and if $B^2 < 0$ then B contains no points.

Given that bivectors now define pairs of points, as opposed to lines, the obvious question is how do we encode lines? Suppose we construct the line through the points $a, b \in \mathbb{R}^n$. A point on the line is then given by
$$x = \lambda a + (1 - \lambda) b. \tag{4.57}$$
The conformal version of this line is
$$\begin{aligned} F(x) &= \left(\lambda^2 a^2 + 2\lambda(1-\lambda) a \cdot b + (1-\lambda)^2 b\right) n + 2\lambda a + 2(1-\lambda) b - \bar{n} \\ &= \lambda A + (1-\lambda) B + \tfrac{1}{2}\lambda(1-\lambda) A \cdot B\, n, \end{aligned} \tag{4.58}$$
and any multiple of this encodes the same point on the line. It is clear, then, that a conformal point X is a linear combination of A, B and n, subject to the constraint that $X^2 = 0$. This is summarised by
$$(A \wedge B \wedge n) \wedge X = 0, \quad X^2 = 0. \tag{4.59}$$
So it is the *trivector* $A \wedge B \wedge n$ which represents a line in conformal geometry. This illustrates a general feature of the conformal model — geometric objects are represented by multivectors one grade higher than their projective

counterpart. The extra degree of freedom is absorbed by the constraint that $X^2 = 0$.

Now suppose that we form a general trivector L from three null vectors,
$$L = A_1 \wedge A_2 \wedge A_3. \tag{4.60}$$
This must still encode a conformal line via the equation $L \wedge X = 0$. In fact, L encodes the *circle* through the points defined by A_1, A_2 and A_3. To see why, consider the conformal model of a plane. The trivector L therefore maps to a *dual* vector l, where
$$l = IL, \tag{4.61}$$
and I is the pseudoscalar defined in equation (4.28). We see that
$$l^2 = L^2 = -2(A_1 \cdot A_2)(A_1 \cdot A_3)(A_2 \cdot A_3), \tag{4.62}$$
so l is a vector with positive square. Such a vector can always be written in the form
$$l = \lambda(F(c) - \rho^2 n) = \lambda(C - \rho^2 n), \tag{4.63}$$
where $C = F(c)$ is the conformal vector for the point $c \in \mathbb{R}^n$. The dual version of the equation $X \wedge L = 0$ is
$$X \cdot l = 0, \tag{4.64}$$
which reduces to
$$\frac{X \cdot C}{X \cdot n} = \rho^2. \tag{4.65}$$
Since $C = F(c)$ satisfies $C \cdot n = -2$, equation (4.65) states that the Euclidean distance between x and c is equal to the constant ρ. This clearly defines a circle in a plane. Furthermore, the radius of the circle is defined by
$$\rho^2 = \frac{l^2}{(l \cdot n)^2} = -\frac{L^2}{(L \wedge n)^2}, \tag{4.66}$$
where the minus sign in the final expression is due to the fact that the 4-vector $L \wedge n$ has negative square. This equation demonstrates how the conformal framework allows us to encode dimensional concepts such as radius while keeping multivectors like L as homogeneous representations of geometric objects.

The equation for the radius of the circle tells us that the circle has infinite radius if
$$L \wedge n = 0. \tag{4.67}$$
This is the case of a straight line, and this equation can be interpreted as saying the line passes through the point at infinity. So, given three points, the test that they lie on a line is
$$A \wedge B \wedge C \wedge n = 0 \implies A, B, C \text{ collinear}. \tag{4.68}$$

This test has an important advantage over the equivalent test in projective geometry. The degree to which the right-hand side differs from zero directly measures how far the points are from lying on a common line. This can resolve a range of problems caused by the finite numerical precision of most computer algorithms. Numerical drift can only affect how near to being straight the line is. When it comes to plotting, one can simply decide what tolerance is required and modify equation (4.68) to read

$$-\frac{(L \wedge n)^2}{L^2} < \epsilon \implies \text{sufficiently straight,} \tag{4.69}$$

where $L = A \wedge B \wedge C$. A similar idea cannot be applied so straightforwardly in the projective framework, as there is no intrinsic measure of being 'nearly linearly dependent' for projective vectors.

Next, suppose that a and b define two vectors in $\mathbb{R}^n$, both as rays from the origin, and we wish to find the angle between these. The conformal representation of the line can be built up from

$$F(a) \wedge F(\lambda a) \wedge F(\mu a) \propto ae\bar{e} = aN, \tag{4.70}$$

where $N = e\bar{e}$ and $N^2 = 1$. Similarly the line in the b direction is represented by bN. We can therefore write

$$L_1 = aN, \qquad L_2 = bN, \tag{4.71}$$

so that

$$L_1 L_2 = ab. \tag{4.72}$$

If we let angle brackets $\langle M \rangle$ denote the scalar part of the multivector M we see that

$$\frac{\langle L_1 L_2 \rangle}{|L_1||L_2|} = \frac{a \cdot b}{|a||b|} = \cos\theta, \tag{4.73}$$

where θ is the angle between the lines. This is true for lines through the origin, but the expression in terms of L_1 and L_2 is unchanged under the action of a general rotor, so applies to lines meeting at any point, and to circles as well as straight lines.

Similar considerations apply to circles and planes in three dimensional space. Suppose that the points a, b, c define a plane in $\mathbb{R}^n$, so that an arbitrary point in the plane is given by

$$x = \alpha a + \beta b + \gamma c, \qquad \alpha + \beta + \gamma = 1. \tag{4.74}$$

The conformal representation of x is

$$X = \alpha A + \beta B + \gamma C + \delta n, \tag{4.75}$$

Conformal Geometry, Euclidean Space and Geometric Algebra　　　　　　　　　　55

where $A = f(a)$ etc., and

$$\delta = \tfrac{1}{2}(\alpha\beta A\cdot B + \alpha\gamma A\cdot C + \beta\gamma B\cdot C). \tag{4.76}$$

Varying α and β, together with the freedom to scale $F(x)$, now produces general null combinations of the vectors A, B, C and n. The equation for the plane can then be written

$$A \wedge B \wedge C \wedge n \wedge X = 0, \qquad X^2 = 0. \tag{4.77}$$

So, as one now expects, it is 4-vectors which define planes. If instead we form the 4-vector

$$S = A_1 \wedge A_2 \wedge A_3 \wedge A_4, \tag{4.78}$$

then in general $S \wedge X = 0$ defines a sphere. To see why, consider the dual of S,

$$s = IS, \tag{4.79}$$

where I is now given by equation (4.31). Again we find that $s^2 > 0$, and we can write

$$s = \lambda(F(c) - \rho^2 n) = \lambda(C - \rho^2 n). \tag{4.80}$$

The equation $X \wedge S = 0$ is now equivalent to $X \cdot s = 0$, which defines a sphere of radius ρ, with centre $C = F(c)$. The radius of the sphere is defined by

$$\rho^2 = \frac{s^2}{(s\cdot n)^2} = \frac{S^2}{(S\wedge n)^2}. \tag{4.81}$$

The sphere becomes a flat plane if the radius is infinite, so the test that four points lie on a common plane is

$$A_1 \wedge A_2 \wedge A_3 \wedge A_4 \wedge n = 0 \quad \Longrightarrow \quad A_1 \ldots A_4 \text{ coplanar}. \tag{4.82}$$

As with the case of the test of collinearity, this is numerically well conditioned, as the deviation from zero is directly related to the curvature of the sphere through the four points.

7.　　Reflection and intersection in conformal space

The conformal model extends the range of geometric primitives beyond the simple lines and planes of projective geometry. It also provides a range of new algorithms for handling intersections and reflections. First, suppose we wish to find the intersection of two lines in a plane. These are defined by the trivectors L_1 and L_2. The intersection, or meet, is defined via its dual by

$$(L_1 \vee L_2)^* = L_1^* \wedge L_2^*. \tag{4.83}$$

The star, L^*, denotes the dual of a multivector, which in geometric algebra is formed by multiplication by the multivector representing the join. For the case of two distinct lines in a plane, their join is the plane itself, so the dual is formed by multiplication by the pseudoscalar I of equation (4.28). The intersection of two lines therefore results in the bivector

$$B = (L_1 \vee L_2)^* = (IL_1) \cdot L_2 = L_1 \cdot (IL_2). \tag{4.84}$$

This expression is a bivector as it is the contraction of a vector and a trivector. As discussed in section 4.6, a bivector can encode zero, one or two points, depending on the sign of its square. Two circles can intersect in two points, for example, if they are sufficiently close to one another. Two straight lines will also intersect in two points, though one of these is at infinity.

In three dimensions a number of further possibilities arise. The intersection of a line L and a sphere or plane P also results in a bivector,

$$L \vee P = L \cdot (IP), \tag{4.85}$$

where I is now the grade-5 pseudoscalar in the conformal algebra for three-dimensional Euclidean space. The bivector encodes the fact that a line can intersect a sphere in up to two places. If P is a flat plane and L a straight line, then one of the intersection points will be at infinity. More complex is the case of two planes or spheres. In this case both geometric primitives are 4-vectors, P_1 and P_2. Their intersection results in the trivector

$$L = P_1 \vee P_2 = (IP_1) \cdot P_2 = P_1 \cdot (IP_2). \tag{4.86}$$

This trivector encodes a line. This one single expression covers a wide range of situations, since either plane can be flat or spherical. If both planes are flat, their intersection is a straight line with $L \wedge n = 0$. If one or both of the planes are spheres, their intersection results in a circle. The sign of the square of the trivector L immediately encodes whether or not the spheres intersect. If $L^2 < 0$ the spheres do not intersect, as there are no null solutions to $L \wedge X = 0$ when $L^2 < 0$.

As well as new intersection algorithms, the application of geometric algebra in conformal space provides a highly compact encoding of reflections. As a single example, suppose that we wish to reflect the line L in the plane P. To start with, suppose that the point of intersection is the origin, with L a straight line in the a direction, and P the plane defined by the points b, c and the origin. In this case we have

$$L = aN, \qquad P = (b \wedge c)N. \tag{4.87}$$

The result of reflecting a in $b \wedge c$ is a vector in the a' direction, where

$$a' = (b \wedge c) a (b \wedge c). \tag{4.88}$$

The conformal representation of the line through the origin in the a' direction is $L' = a'N$. In terms of their conformal representations, we have

$$L' = PLP. \qquad (4.89)$$

So far this is valid at the origin, but conformal invariance ensures that it holds for all intersection points. This expression for L' finds the reflected line in space without even having to find the intersection point. Furthermore, the line can be curved, and the same expression reflects a circle in a plane. We can also consider the case where the plane itself is curved into a sphere, in which case the transformation of equation (4.89) corresponds to an *inversion* in the sphere. Inversions are important operations in geometry, though perhaps of less interest in graphics applications. To find the reflected line for the spherical case is also straightforward, as all one needs is the formula for the tangent plane to the sphere S at the point of intersection. This is

$$P = (X \cdot S) \wedge n, \qquad (4.90)$$

where X is the point where the line L intersects the sphere. The plane P can then be used in equation (4.89) to find the reflected line.

8. Conclusions

Conformal geometry provides an efficient framework for studying Euclidean geometry because the inner product of conformal vectors is directly related to the Euclidean distance between points. The true power of the conformal framework only really emerges when the subject is formulated in the language of geometric algebra. The geometric product unites the outer product, familiar from projective geometry, with the inner product of conformal vectors. Each graded subspace encodes a different geometric primitive, which provides for good data typing and aids writing robust code. Furthermore, the algebra includes multivectors which generate Euclidean transformations. So both geometric objects, and the transformations which act on them, are contained in a single, unified framework.

The remaining step in implementing this algebra in applications is to construct fast algorithms for multiplying multivectors. Much as projective geometry requires a fast engine for multiplying 4×4 matrices, so conformal geometric algebra requires a fast engine for multiplying multivectors. One way to achieve this is to encode each multivector via its matrix representation. For the important case of $\mathcal{G}(4,1)$, this representation consists of 4×4 complex matrices. But in general this approach is slower than algorithms designed to take advantage of the unique properties of geometric algebra. Such algorithms are under development by a number of groups around the world. These will doubtless be of considerable interest to the graphics community.

Acknowledgments

CD would like to thank the organisers of the conference for all their work during the event, and their patience with this contribution. CD is supported by an EPSRC Advanced Fellowship.

References

[1] H. Grassmann. *Die Ausdehnungslehre*. Enslin, Berlin, 1862.

[2] I. Stewart. Hermann Grassmann was right (News and Views). *Nature*, 321:17, 1986.

[3] J. Richter-Gebert and U. Kortenkamp. *The Interactive Geometry Software Cinderella*. Springer, 1999.

[4] D. Hestenes and G. Sobczyk. *Clifford Algebra to Geometric Calculus*. Reidel, Dordrecht, 1984.

[5] C.J.L. Doran and A.N. Lasenby. *Geometric Algebra for Physicists*. Cambridge University Press, 2002.

[6] C.J.L. Doran and A.N. Lasenby. *Physical Applications of Geometric Algebra*. Cambridge Univesity lecture course. Lecture notes available from http://www.mrao.cam.ac.uk/~clifford.

[7] D. Hestenes, H. Li, and A. Rockwood. New algebraic tools for classical geometry. In G. Sommer, editor, *Geometric Computing with Clifford Algebras*. Springer, Berlin, 1999.

[8] D. Hestenes, H. Li, and A. Rockwood. Generalized homogeneous coordinates for computational geometry. In G. Sommer, editor, *Geometric Computing with Clifford Algebras*. Springer, Berlin, 1999.

[9] A.N. Lasenby and J. Lasenby. Surface evolution and representation using geometric algebra. In R. Cippola, A. Martin, editors, *The Mathematics of Surfaces IX: Proceedings of the Ninth IMA Conference on the Mathematics of Surfaces*, pages 144–168. London, 2000.

Chapter 5

TOWARDS THE ROBUST INTERSECTION OF IMPLICIT QUADRICS

Laurent Dupont, Sylvain Lazard, Sylvain Petitjean
Loria-CNRS & INRIA Lorraine
BP 239, 54506 Vandoeuvre cedex, France
{dupont,lazard,petitjea}@loria.fr

Daniel Lazard
LIP6, Universite Pierre et Marie Curie
Boite 168, 4 place Jussieu, 75252 Paris cedex 05, France
Daniel.Lazard@lip6.fr

Abstract   We are interested in efficiently and robustly computing a parametric form of the intersection of two implicit quadrics with rational coefficients. Our method is similar in spirit to the general method introduced by J. Levin for computing an explicit representation of the intersection of two quadrics, but extends it in several directions. Combining results from the theory of quadratic forms, a projective formalism and new theorems characterizing the intersection of two quadratic surfaces, we show how to obtain parametric representations that are both "simple" (the size of the coefficients is small) and "as rational as possible".

Keywords: Robustness of geometric computations, quadric surface intersection

1.   Introduction

Computing the curve of intersection of quadric surfaces is an important step in the boundary evaluation of second-order constructive solid geometry (CSG) solids [13] and in the determination of the convex hull of quadric surface patches [7].

In solid modeling, the two most widely used types of object representation are CSG and boundary representation (B-Rep). Since both representations have their own respective advantages (modeling is more flexible and intuitive with primitive solids, but rendering on graphical display systems is easier with B-Reps), solid modeling kernels often need an efficient and reliable way to switch from CSG to B-Rep representation. Boundary evaluation, also known as CSG-to-B-Rep conversion, is a well understood problem. However, past approaches have often put more emphasis on efficiency than on robustness and accuracy. Most current modelers use only finite-precision arithmetic for CSG-to-B-Rep conversion. The topological consistency of the computed B-Rep can easily be jeopardized by small amounts of error in the data introduced by finite-precision computations. For many applications in design and automated manufacturing, where topological consistency and possibly accuracy are critical, this may be unacceptable.

Designing reliable, accurate and robust algorithms is currently a major interest of the computational geometry and solid modeling research communities (see, e.g., [2, 14]). A number of approaches have been proposed for the robust and accurate boundary evaluation of polyhedral models [1, 4]. Most rely heavily on numerical computation, with varying dependence on exact and floating-point arithmetic. Computing the topological structure of a B-Rep involves accurate evaluation of signs of arithmetic expressions. Assuming the input data has a bounded precision and allowing whatever bit-length is necessary for number representation, these signs can be computed exactly.

By contrast, there has been much less work on robust CSG-to-B-Rep conversion algorithms for curved primitives. A major reason is that outside the linear realm, exact arithmetic computations require algebraic numbers which cannot in general be represented explicitly with a finite number of bits. In addition, computation with algebraic numbers is extremely slow. One notable exception is the work of Keyser et al. [8, 9] on the boundary evaluation of low-degree CSG solids specified with rational parametric surfaces. The authors use exact arithmetic, present compact data structures for representing the boundary curves as algebraic curves and the boundary vertices as algebraic numbers and use efficient algorithms for computing the intersection curves of parametric surfaces.

The quadratic nature of the equations defining quadric surfaces permits an explicit representation of the intersection curves. In other words, it is theoretically possible to compute a fully parametric representation of the boundary of quadric-based solids. The general method for computing an explicit parametric representation of the intersection between two quadrics is due to J. Levin [11, 12]. This seminal work has been extended in many different directions. For instance, arguing that Levin's method does not yield explicit information on the morphological type of the intersection curve, Farouki et al. [3] made a complete theoretical study, for general quadric surfaces, of degenerate cases. Goldman and Miller [6] took a different path and developed a special-case solution for each of the possible pairs of natural quadrics (i.e., planes, right cones, circular cylinders and spheres).

Most of these methods were motivated by the belief that general methods for intersecting implicit quadric surfaces may not be numerically robust and may fail in degenerate configurations. Indeed, no one has reported thus far an algorithm for robustly computing the intersection between two general quadric surfaces. Recently, however, Geismann et al. [5] have shown how to exactly compute a cell in an arrangement of three quadrics.

We show in this paper that the method of Levin can be improved in several ways in order to remove most of the sources of numerical instabilities in the original algorithm. Using a combination of projective formalism, reduction of quadratic forms and new theorems characterizing the intersection of some quadric surfaces, we show, in particular, how to avoid the appearance of nested radicals (high-degree algebraic numbers).

We present in Section 5.2 basic definitions and notation. We recall in Section 5.3 Levin's method for intersecting two quadric surfaces, and present our method in Section 5.4.

2. Preliminaries

In the rest of this paper, we assume that each quadric surface $\mathcal{P}$ is given as the zero-set in $\mathbb{R}^3$ of a quadratic implicit equation P with rational coefficients in the variables (x_1, x_2, x_3), i.e.,

$$\alpha_1 x_1^2 + \alpha_2 x_2^2 + \alpha_3 x_3^2 + 2\alpha_4 x_1 x_2 + 2\alpha_5 x_1 x_3 + 2\alpha_6 x_2 x_3 \\ + 2\alpha_7 x_1 + 2\alpha_8 x_2 + 2\alpha_9 x_3 + \alpha_{10} = 0, \quad (5.1)$$

with $\alpha_i \in \mathbb{Q}, i = 1, \ldots, 10$ and some $\alpha_i \neq 0$. Eq. (5.1) can be rewritten in matrix form as $\mathbf{X}^T P \mathbf{X} = 0$, with $\mathbf{X} = (x_1, x_2, x_3, 1)^T$ and P the symmetric 4×4 matrix

$$P = \begin{pmatrix} \alpha_1 & \alpha_4 & \alpha_5 & \alpha_7 \\ \alpha_4 & \alpha_2 & \alpha_6 & \alpha_8 \\ \alpha_5 & \alpha_6 & \alpha_3 & \alpha_9 \\ \alpha_7 & \alpha_8 & \alpha_9 & \alpha_{10} \end{pmatrix}.$$

We use the same notation P for a quadratic implicit equation and its associated matrix; the corresponding quadric surface is denoted $\mathcal{P}$.

A quadric is said to be given in *canonical form* if its equation in some coordinate frame is of the form

$$\sum_{i=1}^{p} a_i x_i^2 - \sum_{i=p+1}^{r} a_i x_i^2 + \xi = 0,$$

or

$$\sum_{i=1}^{p} a_i x_i^2 - \sum_{i=p+1}^{r} a_i x_i^2 - x_{r+1} = 0, \tag{5.2}$$

with $a_i > 0 \forall i, \xi \in \{0, 1\}$ and $p \leqslant r$.

The signature of P is an ordered pair (p, q) where p and q are the numbers of positive and negative eigenvalues of P, respectively (the number of null eigenvalues thus follows from the signature). If P has signature (p, q) then $-P$ has signature (q, p), but the quadric surfaces associated with P and $-P$ are identical. We thus define here the *signature* of a quadric surface $\mathcal{P}$ as the couple (p, q), $p \geqslant q$, where p and q are the numbers of positive and negative eigenvalues of P (non-respectively).

We will refer to the 3×3 upper left submatrix of P, denoted P_u, as the *principal submatrix* of P. The determinant of this matrix is called the *principal subdeterminant*.

Given two quadrics $\mathcal{P}$ and $\mathcal{Q}$, the *pencil* generated by $\mathcal{P}$ and $\mathcal{Q}$ is the set of quadrics $\mathcal{R}(\lambda, \mu)$ of equation $\lambda P + \mu Q$ where $(\lambda, \mu) \in \mathbb{R}^2 \setminus \{0, 0\}$. For simplicity of notation, we will consider instead the pencil generated by $\mathcal{P}$ and $\mathcal{Q}$ as the set of quadrics $\mathcal{R}(\lambda)$ of the equation $P - \lambda Q$, $\lambda \in \mathbb{R}$, augmented by the matrix Q. For simplicity of presentation, we will write that the quadric $\mathcal{R}(\lambda)$ is equal to $\mathcal{P} - \lambda \mathcal{Q}$. Finally, recall the well-known result that the intersection between two distinct quadric surfaces in a pencil does not depend on the choice of the two quadrics.

3. Levin's method

We describe in this section the method presented by J. Levin in [11, 12] for computing a parameterized expression for the intersection of two quadric surfaces given by their implicit equations.

Assume that $\mathcal{P}$ and $\mathcal{Q}$ are two distinct quadric surfaces. Levin's method is based on his following key result.

THEOREM 5.1 ([11]) *The pencil generated by two distinct quadric surfaces contains at least one simple ruled quadric, i.e., the empty set or a quadric listed in Table 5.1.*

One important property of simple ruled quadrics is that the value of their principal sub-determinants is zero. Levin's method is as follows.

1 Find a simple ruled quadric in the pencil generated by $\mathcal{P}$ and $\mathcal{Q}$. This is achieved by computing the type of $\mathcal{Q}$ and of the quadrics $\mathcal{R}(\lambda) = \mathcal{P} - \lambda \mathcal{Q}$ such that λ is solution[1] of $\det(R_u(\lambda)) = 0$. By Theorem 5.1, one of these quadric surfaces is simple ruled. Let $\mathcal{R}$ be such a quadric and assume that $\mathcal{R}$ and $\mathcal{P}$ are distinct (otherwise, choose $\mathcal{Q}$ instead of $\mathcal{P}$).

Table 5.1. Parameterization of canonical simple ruled quadrics.

quadric	canonical equation $a_i > 0$	parameterization $\mathbf{X} = [x_1, x_2, x_3], u, v \in \mathbb{R}$
line	$a_1 x_1^2 + a_2 x_2^2 = 0$	$\mathbf{X}(u) = [0, 0, u]$
simple plane	$x_1 = 0$	$\mathbf{X}(u, v) = [0, u, v]$
double plane	$a_1 x_1^2 = 0$	$\mathbf{X}(u, v) = [0, u, v]$
parallel planes	$a_1 x_1^2 = 1$	$\mathbf{X}(u, v) = [\frac{1}{\sqrt{a_1}}, u, v]$, $\mathbf{X}(u, v) = [-\frac{1}{\sqrt{a_1}}, u, v]$
intersecting planes	$a_1 x_1^2 - a_2 x_2^2 = 0$	$\mathbf{X}(u, v) = [\frac{u}{\sqrt{a_1}}, \frac{u}{\sqrt{a_2}}, v]$, $\mathbf{X}(u, v) = [\frac{u}{\sqrt{a_1}}, -\frac{u}{\sqrt{a_2}}, v]$
hyperbolic paraboloid	$a_1 x_1^2 - a_2 x_2^2 - x_3 = 0$	$\mathbf{X}(u, v) = [\frac{u+v}{2\sqrt{a_1}}, \frac{u-v}{2\sqrt{a_2}}, uv]$
parabolic cylinder	$a_1 x_1^2 - x_2 = 0$	$\mathbf{X}(u, v) = [u, a_1 u^2, v]$
hyperbolic cylinder	$a_1 x_1^2 - a_2 x_2^2 = 1$	$\mathbf{X}(u, v) = [\frac{1}{2\sqrt{a_1}}(u + \frac{1}{u}), \frac{1}{2\sqrt{a_2}}(u + \frac{1}{u}), v]$

2 Compute the orthogonal transformation $\mathcal{T}$ which sends $\mathcal{R}$ into canonical form. In the orthonormal coordinate frame in which $\mathcal{R}$ is canonical, $\mathcal{R}$ has one of the parameterizations $\mathbf{X}$ of Table 5.1. Compute the matrix $P' = \mathcal{T}^{-1} P \mathcal{T}$ of the quadric surface $\mathcal{P}$ in the canonical frame of $\mathcal{R}$ and consider the equation (augment $\mathbf{X}$ by a fourth coordinate set to 1)

$$\mathbf{X}^T P' \mathbf{X} = a(u)v^2 + b(u)v + c(u) = 0. \tag{5.3}$$

The parameterizations of Table 5.1 are such that $a(u), b(u)$ and $c(u)$ are polynomials of degree two in u.

3 Solve (5.3) for v in terms of u and determine the corresponding domain of validity of u on which the solutions are defined. This domain can be computed exactly since it is the set of u for which the polynomial of degree four $\Delta(u) = b^2(u) - 4a(u)c(u) \geq 0$. Substituting v by its expression in terms of u in $\mathbf{X}$, we have a parameterization of $\mathcal{P} \cap \mathcal{Q} = \mathcal{P} \cap \mathcal{R}$ in the orthonormal coordinate system where $\mathcal{R}$ is canonical.

4 Report $\mathcal{T}\mathbf{X}(u)$ and the corresponding domain of $u \in \mathbb{R}$ on which the solution is defined, as the parameterized equation of $\mathcal{P} \cap \mathcal{Q}$ in the global coordinate frame.

Levin's method is very nice and powerful since it gives an explicit parametric expression for the intersection of two quadric surfaces given by their implicit equations. However, in terms of precision and robustness, the method is not ideal because it introduces many irrational numbers. Thus, if a floating point representation of numbers is used, the result is imprecise or worse, the implementation crashes if no simple ruled quadric is found in step one. If an exact arithmetic representation is used, then the computations are slow because of the high degree of the algebraic numbers; in practice however, because of the high degree of these algebraic numbers, a correct implementation using exact arithmetic seems fairly impossible.

To be precise, the method introduces irrational numbers (namely square and cubic roots) at the following different steps:

- In Step 1, λ is the root of a degree 3 polynomial in $\mathbb{Q}[X]$; λ can thus be expressed with two levels of nested radicals (cubic and square roots).
- In Step 2, the coefficients of the transformation matrix $\mathcal{T}$ are expressed with 4 levels of nested roots; indeed, the eigenvalues of R_u are the roots of a degree-two polynomial in $\mathbb{Q}[X, \lambda]$ (the degree of the polynomial is two instead of three because $\mathcal{R}$ is a simple ruled quadric); furthermore, the eigenvectors have to be normalized, introducing a new level in the nested square roots; the coefficients of the transformation matrix are thus expressed with 2 levels of nested square roots in $\mathbb{Q}[\lambda]$. Also in Step 2, some other coefficients expressed with 4 levels of nested roots appear in the parameterization $\mathbf{X}$. Indeed, these terms appear in $\mathbf{X}$ as the square roots of the eigenvalues of R_u, which are solutions of a degree-two polynomial in $\mathbb{Q}[X, \lambda]$.
- In Step 3, a square root appears when solving the degree 2 equation (5.3). Note, however, that this radical is the square root of a polynomial in u, the parameter of the curve of intersection.

It is interesting to point out that the solutions produced by Levin's method tend to be very complicated in practice. Consider the simple example where $\mathcal{P}$ is a hyperbolic paraboloid with equation $ax_1^2 + bx_1x_2 - x_3 + 1 = 0, a, b > 0$ and $\mathcal{Q}$ is anything. Then, $\mathcal{R}$ can be chosen equal to $\mathcal{P}$, which substantially simplifies the solution ($\lambda = 0$ and the number of levels of nested radical is 2 instead of 4). Even then, the solution $\mathcal{T}\mathbf{X}$ has the form

$$\begin{pmatrix} \frac{b(u+v)/\sqrt{2}}{\sqrt{b^2+(\sqrt{a^2+b^2}-a)^2}\sqrt{\sqrt{a^2+b^2}+a}} + \frac{b(u-v)/\sqrt{2}}{\sqrt{b^2+(\sqrt{a^2+b^2}+a)^2}\sqrt{\sqrt{a^2+b^2}-a}} \\ \frac{(\sqrt{a^2+b^2}-a)(u+v)/\sqrt{2}}{\sqrt{b^2+(\sqrt{a^2+b^2}-a)^2}\sqrt{\sqrt{a^2+b^2}+a}} - \frac{(\sqrt{a^2+b^2}+a)(u-v)/\sqrt{2}}{\sqrt{b^2+(\sqrt{a^2+b^2}+a)^2}\sqrt{\sqrt{a^2+b^2}-a}} \\ uv+1 \\ 1 \end{pmatrix}.$$

where v still has to be replaced by its solution of (5.3) in terms of u.

4. Our method

As in Levin's method, we want to compute an explicit parametric expression for the intersection of two quadric surfaces given by their implicit equations. However, we only report the components of the intersection that are of strictly positive dimension. This choice is consistent with the applications we are interested in, such as CSG-to-B-Rep conversion, because it only means that the primitives defining the CSG models are considered as open volumes. As Theorem 5.3 will show, not reporting components of dimension zero might simplify the computations.

We improve Levin's method in the following different ways. First we consider the quadric surfaces in the real projective space of dimension three, $\mathbb{P}^3$. One of the keys to Levin's method is the existence of parameterizations of "enough" canonical quadric surfaces (the simple ruled quadrics) such that after substitution into the implicit equation of another quadric, we get a second degree equation in one variable with a discriminant Δ of degree four in another variable. We generalize to projective space that aspect of Levin's method by presenting in Table 5.2 parameterizations, with the above property, for all quadrics, except those of signature $(3, 1)$ (see Theorem 5.2). (Euclidean quadrics of signature $(3, 1)$ are ellipsoids, hyperboloids of two sheets, and elliptic paraboloids.) Considering quadrics in projective space instead of Euclidean space reduces the number of times a quadric $\mathcal{R}$, distinct from $\mathcal{P}$, $\mathcal{Q}$, has to be searched for in the pencil; indeed, in projective space, a search has to be performed if and only if both $\mathcal{P}$ and $\mathcal{Q}$ are of signature $(3, 1)$, although, in Euclidean space, a search has to be performed if and only if both $\mathcal{P}$ and $\mathcal{Q}$ are of signature $(3, 1)$ or amongst the cones, elliptic cylinders or hyperboloids of one sheet.

Table 5.2. Parameterization of projective quadric surfaces.

signature of $\mathcal{Q}$	canonical equation $a_i > 0$	parameterization $\mathbf{X} = [x_1, x_2, x_3, x_4] \in \mathbb{P}^3$
$(4, 0)$	$a_1 x_1^2 + a_2 x_2^2 + a_3 x_3^2 + a_4 x_4^2 = 0$	$\mathcal{Q} = \emptyset$
$(3, 1)$	$a_1 x_1^2 + a_2 x_2^2 + a_3 x_3^2 - a_4 x_4^2 = 0$	$\det Q < 0$
$(3, 0)$	$a_1 x_1^2 + a_2 x_2^2 + a_3 x_3^2 = 0$	$\mathcal{Q}$ is a point
$(2, 2)$	$a_1 x_1^2 + a_2 x_2^2 - a_3 x_3^2 - a_4 x_4^2 = 0$	$\mathbf{X} = [\frac{ut+vw}{\sqrt{a_1}}, \frac{uw-vt}{\sqrt{a_2}}, \frac{ut-vw}{\sqrt{a_3}}, \frac{uw+vt}{\sqrt{a_4}}]$, $(u,v),(w,t) \in \mathbb{P}^1$
$(2, 1)$	$a_1 x_1^2 + a_2 x_2^2 - a_3 x_3^2 = 0$	$\mathbf{X} = [\frac{u^2+v^2}{2\sqrt{a_1}}, \frac{u^2-v^2}{2\sqrt{a_2}}, \frac{uv}{\sqrt{a_3}}, wt]$, $(u,v,w,t) \in \mathbb{P}^3$
$(2, 0)$	$a_1 x_1^2 + a_2 x_2^2 = 0$	$\mathbf{X} = [0, 0, u, v], (u, v) \in \mathbb{P}^1$
$(1, 1)$	$a_1 x_1^2 - a_2 x_2^2 = 0$	$\mathbf{X} = [\frac{u}{\sqrt{a_1}}, \frac{u}{\sqrt{a_2}}, v, w]$, $\mathbf{X} = [\frac{u}{\sqrt{a_1}}, -\frac{u}{\sqrt{a_2}}, v, w]$, $(u,v,w) \in \mathbb{P}^2$
$(1, 0)$	$a_1 x_1^2 = 0$	$\mathbf{X} = [0, u, v, w], (u, v, w) \in \mathbb{P}^2$

THEOREM 5.2 *Let P' be a symmetric real 4×4 matrix. The parameterizations $\mathbf{X}$ presented in Table 5.2 are such that $\mathbf{X}^T P' \mathbf{X}$ is equal to a $w^2 + b\,wt + c\,t^2$, $a\,(wt)^2 + b\,wt + c$, or $a\,w^2 + b\,w + c$, for the first, second and three last parameterizations respectively, where $\Delta = b^2 - 4ac$ is a homogeneous polynomial of degree 4 in the variables u, v.*

Our other key result for avoiding the appearance of irrational numbers is the following:

THEOREM 5.3 *If two quadric surfaces $\mathcal{P}$ and $\mathcal{Q}$ both have signature $(3,1)$ and intersect in more than two points (in $\mathbb{P}^3$) then there exists a rational number λ such that $\mathcal{P} - \lambda\mathcal{Q}$ is not of signature $(3,1)$.*

This result is of interest because of the following two reasons. First it ensures that the two quadrics, $\mathcal{P}$ and $\mathcal{R}$, we end up intersecting have rational coefficients. Secondly, the $\lambda \in \mathbb{Q}$, such that $\mathcal{R}(\lambda) = \mathcal{P} - \lambda\mathcal{Q}$ is not of signature $(3,1)$, can be computed, most of the time, with normal floating-point arithmetic. Indeed, first notice that $\mathcal{R}(\lambda)$ is not of signature $(3,1)$ if and only if $\det R(\lambda) \geqslant 0$, and that $\exists \lambda \in \mathbb{R}$ such that $\det R(\lambda) \geqslant 0$, by Theorem 5.1. It follows that, in most of the cases, there exists an interval of $\lambda \in \mathbb{R}$ on which $\det R(\lambda) \geqslant 0$. Thus, we can compute approximatively the roots of $\det R(\lambda) = 0$, then choose a rational number in each interval induced by these computed roots, and find out whether one of these rational numbers is such that $\det R(\lambda) \geqslant 0$. If we found such a λ we report the quadric $\mathcal{R}(\lambda)$ which is not of signature $(3,1)$; due to the lack of space, we do not discuss here how to compute λ otherwise.

Our last contribution is the introduction of Gauss' reduction method for quadratic forms for transforming $\mathcal{R}$ into a canonical frame in Step 2. Using Gauss' reduction method instead of finding an orthogonal transformation as in Levin's method, simplifies substantially the form of the solutions. As an example, using just Gauss' reduction method on the example of Section 5.3 where $\mathcal{P}$ is a hyperbolic paraboloid of equation $ax_1^2 + bx_1x_2 - x_3 + 1 = 0$, $a, b > 0$, gives a solution $\mathcal{T}\mathbf{X}$ of the form $(\frac{v}{\sqrt{a}}, \frac{\sqrt{a}(u-v)}{b}, uv+1, 1)^T$.

Our algorithm is as follows.

1 Find a quadric surface $\mathcal{R}$ with rational coefficients in the pencil generated by $\mathcal{P}$ and $\mathcal{Q}$ such that $\det R \geqslant 0$, or report that the intersection of $\mathcal{P}$ and $\mathcal{Q}$ is of dimension 0. This is achieved as follows. If $\det Q \geqslant 0$ then choose $\mathcal{R} = \mathcal{Q}$. Otherwise, if there exists $\lambda \in \mathbb{Q}$ such that $\det(P - \lambda Q) \geqslant 0$, then compute such a λ and set $\mathcal{R} = \mathcal{P} - \lambda\mathcal{Q}$. Otherwise, report that $\mathcal{P} \cap \mathcal{Q}$ is of dimension 0 (see Theorem 5.3).

2 If $\mathcal{R}$ has signature $(4, 0)$ or $(3, 0)$ then report that $\mathcal{P} \cap \mathcal{Q}$ is of dimension 0 (see Table 5.2). Otherwise, using Gauss' method for reducing quadratic forms into diagonal forms [10], compute a (non-orthogonal) transformation $\mathcal{T}$ which sends R into canonical form. In that coordinate frame, in which $\mathcal{T}^T R \mathcal{T}$ is diagonal, the quadric surface $\mathcal{R}$ has one of the parameterizations $\mathbf{X}$ of Table 5.2 (by Sylvester's Inertia Law, the signatures of R and $\mathcal{T}^T R \mathcal{T}$ are equal). Compute the matrix $P' = \mathcal{T}^T P \mathcal{T}$ of the quadric surface $\mathcal{P}$ in that coordinate frame and consider the equation $\mathbf{X}^T P' \mathbf{X} = 0$.

3. Solve $\mathbf{X}^T P' \mathbf{X} = 0$ as an equation of degree two according to Theorem 5.2, and determine the domains of $(u, v) \in \mathbb{P}^1$ on which the solutions are defined. Substitute, depending on the case, (w, t), wt or w in terms of (u, v) in $\mathbf{X}$. (If a, b, c vanish simultaneously for some values of (u, v), then replace (u, v) by those values in $\mathbf{X}$ and let $(w, t) \in \mathbb{P}^1$.) We get a parameterization of $\mathcal{P} \cap \mathcal{Q} = \mathcal{P} \cap \mathcal{R}$ in the coordinate system where $\mathcal{R}$ is canonical.
4. Report $\mathcal{T}\mathbf{X}(u, v)$ and the corresponding domain of $(u, v) \in \mathbb{P}^1$ on which the solution is defined, as the parameterized equation of $\mathcal{P} \cap \mathcal{Q}$ in the global coordinate frame.

The algorithm we presented here produces an explicit parametric representation for the intersection of two quadric surfaces $\mathcal{P}$ and $\mathcal{Q}$ given by their implicit equations (with rational coefficients), such that all the coefficients are in $\mathbb{Q}[\sqrt{a_1}, \sqrt{a_2}, \sqrt{a_3}, \sqrt{a_4}]$ where the a_i are the coefficients of $\mathcal{R}$ in canonical form through $\mathcal{T}$ (i.e., the coefficients on the diagonal of $\mathcal{T}^T R \mathcal{T}$).

Notes

1. Levin's method fails if $\det(R_u(\lambda)) \equiv 0$ because, then, the type of all the quadrics $\mathcal{R}(\lambda)$, $\lambda \in \mathbb{R}$, have to be computed. However, the method can be easily fixed by considering the roots of other polynomials, such as $\det(R(\lambda))$, that naturally appear when considering the classification of quadric surfaces as in [12].

References

[1] M. Benouamer, D. Michelucci, and B. Peroche. Error-free boundary evaluation based on a lazy rational arithmetic: a detailed implementation. *Computer-Aided Design*, 26(6):403–416, 1994.

[2] A. Bowyer, J. Berchtold, D. Eisenthal, I. Voiculescu, and K. Wise. Interval methods in geometric modeling. In *Proc. of International Conference on Geometric Modeling and Processing, Hong Kong*, 2000. Invited presentation.

[3] R. Farouki, C. Neff, and M. O'Connor. Automatic parsing of degenerate quadric-surface intersections. *ACM Transactions on Graphics*, 8(3):174–203, 1989.

[4] S. Fortune. Polyhedral modelling with exact arithmetic. In *Proc. of ACM Symposium on Solid Modeling and Applications*, pages 225–234, 1995.

[5] N. Geismann, M. Hemmer, and E. Schoemer. Computing the intersection of quadrics: exactly and actually. In *Proc. of ACM Symposium on Computational Geometry*, pages 264–273, 2001.

[6] R. Goldman and J. Miller. Combining algebraic rigor with geometric robustness for the detection and calculation of conic sections in the intersection of two natural quadric surfaces. In *Proc. of ACM Symposium on Solid Modeling Foundations and CAD/CAM Applications*, pages 221–231, 1991.

[7] C.-K. Hung and D. Ierardi. Constructing convex hulls of quadratic surface patches. In *Proceedings of 7th CCCG (Canadian Conference on Computational Geometry), Quebec, Canada*, pages 255–260, 1995.

[8] J. Keyser, S. Krishnan, and D. Manocha. Efficient and accurate B-Rep generation of low degree sculptured solids using exact arithmetic: I – Representations. *Computer Aided Geometric Design*, 16(9):841–859, 1999.

[9] J. Keyser, S. Krishnan, and D. Manocha. Efficient and accurate B-Rep generation of low degree sculptured solids using exact arithmetic: II – Computation. *Computer Aided Geometric Design*, 16(9):861–882, 1999.

[10] T. Lam. *The Algebraic Theory of Quadratic Forms*. W.A. Benjamin, Reading, MA, 1973.

[11] J. Levin. A parametric algorithm for drawing pictures of solid objects composed of quadric surfaces. *Communications of the ACM*, 19(10):555–563, 1976.

[12] J. Levin. Mathematical models for determining the intersections of quadric surfaces. *Computer Graphics and Image Processing*, 11(1):73–87, 1979.

[13] R. Sarraga. Algebraic methods for intersections of quadric surfaces in GMSOLID. *Computer Vision, Graphics, and Image Processing*, 22:222–238, 1983.

[14] K. Sugihara. How to make geometric algorithms robust. *IEICE Transactions on Information and Systems*, E83-D(3):447–454, 2000.

Chapter 6

COMPUTATIONAL GEOMETRY AND UNCERTAINTY

A.R. Forrest
School of Information Systems
University of East Anglia
Norwich NR4 7TJ
United Kingdom
forrest@sys.uea.ac.uk

Abstract	From the prehistory of computational geometry it has been apparent that geometric computation is fraught with problems. Although these problems have become less troublesome over the ensuing thirty years, they have not been eliminated. The paper discusses the sources of geometric errors in applied computational geometry systems and reviews various attempts at eliminating them in practical systems. No completely satisfactory solution has been devised, but for some restricted cases, there has been progress. A possible way ahead which may enable provably correct systems to be implemented is suggested.

1. Introduction

One of the earliest and most important applications of computer graphics was in computer-aided design and in particular geometric modelling. Graphics provided an interface to systems in which geometry was paramount. The goal was to build a computer model which embodied the geometric truth: Do parts fit together ? Do they function correctly ? Can they be manufactured ? The increasing sophistication of designs, particularly the use of complex forms, required new methods of design and manufacture difficult or impossible to achieve with conventional drafting and machining techniques. Aircraft, ships and cars necessitated systems which could handle the shapes dictated by aerodynamics, hydrodynamics or styling. The disciplines involved included differential geometry, approximation theory, and numerical analysis, with computer graphics used to provide user-friendly interfaces which overcame the limitations of conventional drafting.

A primary goal of the early work was to develop geometric methods which could model the required shapes, were amenable to computational techniques, and could provide suitable design handles for users. The resulting technology includes Bezier, B-spline, and NURBS curves and surfaces, and subdivision surfaces which are now widely accepted. Users of such systems were generally highly trained and design tended to proceed by the manipulation of curve and surface boundaries. Models were assembled by stitching bounded regions together with pre-defined continuity. Whilst there were numerical problems, difficulties could generally be overcome by the user's knowledge of how they might be resolved.

When work started on solid modelling at Cambridge University in the late 1960's, the goal was to build systems for what were termed "simple mechanical parts". It was thought that this involved complex assemblies of relatively straightforward shapes bounded by planes, cylinders, etc. and that significant geometric problems would not arise. The design metaphor was based on assembling objects from simple parts by operations akin to machining, welding, etc.: joining objects, drilling holes, and intersecting to create recesses and protrusions, rather than on the creation of complex surfaces. The technologies developed included spatial enumeration, primitive instancing, constructive solid geometry (CSG), and boundary representation (B-rep). Almost from the outset it was realised that simple mechanical parts were not that simple to implement and that there were issues involving robustness, efficiency, and the detection and handling of special cases which had not arisen in curve and surface modelling: in short, we were faced with geometric uncertainty.

2. Why computational geometry ?

Computational geometry was originally defined as "the computer representation, analysis and synthesis of shape information" [Forrest, 1971] and was prompted by issues arising from curve and surface work and experiences with solid modelling. In the case of curves and surfaces, conventional geometry did not address the problems which needed to be solved, both for representation of shapes and for computing with shape information; in the case of solid modelling, an algorithmic approach to geometry was required which necessitated the development of computational tools to replace conventional drafting tools and geometric constructions. Issues of efficiency and computational complexity were regarded as being of lesser importance than the ability to model the required shapes. Independently, around 1974, Shamos at Yale began to work on computational geometry, concentrating on algorithms for 2-D problems involving points and lines, and in particular on complexity [Shamos, 1974, Shamos, 1975, Shamos, 1978]. Shamos posed the question: what are the geometric primitives required for computational geometry? The Euclidean primitives, straight edge and compass, do not necessarily match digital computation, nor do they necessarily lead to efficient, implementable, algorithms.

For some time the two strands of computational geometry developed in parallel and to a large extent independently but in recent years there has been a realisation that they are two sides of the same coin. The annual Computational Geometry Symposium now includes theoretical and applied streams. Theoretical computational geometry is rooted in the theory of algorithms, emphasising complexity analysis, theorems and proofs, design of algorithms and the development of data structures. Assumptions are commonly made regarding input to avoid issues giving rise to special cases: geometric objects are assumed to be in general position, e.g. no parallel lines, no horizontal lines, no more than two co-linear points, no more than three co-circular points, and so on. A welcome trend in recent years has been the absence of such restrictions. By contrast, in geometric modelling and computer-aided design, co-linearity, parallelism, and similar awkward configurations are commonly employed in the construction and representation of objects, and a system must handle all the special cases which arise. Systems must be robust, flexible, and predictable: uncertainty should be eliminated if possible. Computational geometry is therefore concerned with all the issues which arise when solving geometric problems by computer.

3. Sources of uncertainty

Geometric uncertainty arises where computation is performed between two or more geometric entities and is due to two causes: special cases and numerical problems. When computer-aided design was largely concerned with curves and surfaces, design proceeded in many cases by designing the boundary curves of surface patches, so that surface patches intersected at the designed boundaries, rather than by designing surface patches and then trimming the patches by intersection with other patches. Solid modelling, on the other hand, created models by intersection and by attaching objects. Geometric special cases such as parallelism, coincidence, and tangency are an immediate result of this paradigm, and numerical problems follow.

Braid's first solid modeller, BUILD [Braid, 1973], revealed a plethora of geometric special cases which needed special treatment and careful coding. Subsequent modellers uncovered hitherto unimplemented special cases, and it was apparently difficult to prove that a system was capable of handling all the possible special cases which might arise. ROMULUS, the first commercial modeller from the Cambridge stable, failed when one user wished to define a torus tangential to the apex of a cone (a seemingly unlikely operation but employed as a construction step immediately followed by removal of the apex of the cone).

Numerical problems in computational geometry arise from both rounding errors and numerical instabilities. Whereas in the design of algorithms, the input data set generally passes through the algorithm and little data is recycled or passed on to other algorithms, in a design situation, a data structure is created in an incremental fashion with subsequent design iterations giving rise to the accumulation of rounding errors. Even simple geometric operations can give rise to severe numerical problems. For example, the intersection of two lines could be tackled by solving the two linear equations, but the numerical stability of the inherent matrix inversion depends on how orthogonal the matrix is. In geometric terms this simply means that accuracy is highest when lines are perpendicular and lowest when lines are nearly parallel. Given the common occurrence of parallel lines in many applications, and the presence of rounding errors, it can readily be seen that unless systems are carefully implemented, attempts might be made to intersect lines which are essentially parallel. Solomon's 1985 thesis at Cambridge [Solomon, 1985] declared that "it is best to carry out all geometric operations as close to the origin as possible". This is a veiled comment on the inadvisability of subtracting two large floating point numbers and expecting precision in the result but it also points up an insidious form of uncertainty: an operation in a modelling system might yield different outcomes depending on where in the model space the operation is performed.

A second form of uncertainty caused by numerical problems arises from the distinction in computational geometry between detection and computation [Chazelle and Dobkin, 1980]. Detection determines whether a condition exists, whilst computation evaluates the actual condition. For example, line segment intersection detection determines whether two line segments intersect, whilst computation evaluates the point of intersection. Generally speaking detection is easier than computation, so it is conventional to detect first and only evaluate when it is necessary to do so. Unfortunately, the two operations generally involve different chains of numerical operations and it is quite possible for a detection operation to indicate that, say, an intersection exists, but for the computation step to declare that no such point exists, or vice versa: which of the two operations are we to believe?

In 1980 the BUILD solid modeller was ported from an IBM 370 to a PRIME 400 at UEA. This necessitated porting an Algol68 compiler and, in anticipation of numerical problems, the default single precision floating point word length was set to 64 bits. The first simple attempt at creating a model failed due to numerical overflow, but creating the same model by a different chain of operations succeeded! One of the main objections to BUILD and other boundary representation modellers was that whilst they created legal data structures for the faces, edges and vertices of complex models by detecting intersections, etc., there was no guarantee that the numerical values computed to populate the data structures did indeed define legal objects. Boundary representation modellers employ validation procedures in an attempt to remove potential inconsistencies. Constructive solid geometry was promoted as a safer form of solid modelling since the data structure is always provably legal. CSG also maps rather directly the intersection-union-difference user interface normally used in solid modelling. However, the CSG data structure does not contain explicit information regarding faces, edges and vertices – it is implicit and unevaluated, and any attempt to determine boundary information immediately invokes all the numerical problems which afflict boundary representation. After more than thirty years of developing and building solid modellers, including the ACIS modelling kernel, special cases and numerical problems have not been completely eliminated [Braid, 2001].

4. Methods for eliminating uncertainty

An excellent discussion of some of the measures proposed is given by Seidel [Seidel, 1998].

As a first measure to overcome numerical problems, the tactic of increasing floating point precision is often employed. This can reduce the number of instances where problems arise, but as noted earlier, numerical instability will not go away, and sometimes is manifested in apparently simple situations. Rational arithmetic has been employed (as far back is Sketchpad in 1963), but tends to be unwieldy. Some common geometric objects such as circles can give rise to irrational numbers. One solid modeller using rational arithmetic suffered from inefficiencies due to numerators and denominators growing rapidly when it had been expected that common factors would have allowed cancellations. Use of extended or infinite precision arithmetic has also been advocated. Interval arithmetic caters explicitly for error bounds, but intervals can increase to unacceptable sizes where design proceeds incrementally. Algebraic manipulation is sometimes suggested and has a certain appeal. However, if models are constructed entirely in symbolic terms, the exact nature of the model may not be apparent due to conditional configurations (if A intersects B then we have one model, otherwise we have a different model). Combinatorial explosion can occur, and resolution requires, in the end, numerical evaluation of some sort, as remarked earlier in the case of CSG.

Geometric special cases can be handled either by implementing code to handle all cases (assuming they can be identified), or by preprocessing the input data to ensure that no special cases arise. Catering for all special cases is tedious and hard to get correct in a robust fashion. Altering the input data, generally by some form of perturbation, makes algorithm implementation simpler provided the configurations which give rise to special cases can be identified and removed. We must devise some automatic process to perturb the geometry by infinitesimal amounts, solve the resulting different but related problem and then translate back to the original problem. The process is difficult to automate, the complexity of the solution can alter, and translating back to the original problem may be more complex than solving the original problem. The methods commonly rely on and assume the availability of exact arithmetic. If translating back to the original problem proves too difficult, then we cannot be certain as to how valid the solution is.

5. A possible road ahead

As we have seen there are fundamental problems in practical computational geometry: special cases, far from being rare, are commonly found, normal geometric configurations give rise to numerical ill-conditioning, design is incremental, output is data structures rather than datasets, data structures may be the input for further computation, and so on. Proof of completeness and/or correctness is hard if not impossible to supply for implemented as opposed to theoretical systems. Furthermore, some common shapes cannot be represented explicitly on digital machines. The user model of what the system does is generally based on real-world experiences. The standard approach to algorithm development and implementation is based on the models of Euclidean/Cartesian and differential geometry. This mathematical model is then translated into computational models and algorithms with little thought on the consequences of moving from continuous space to the finite discrete space of the computer. At worst, we simply use floating point as equivalent to real arithmetic and ignore the mismatch. The transition from continuous space to discrete space is not well controlled.

Are we using the appropriate geometry for digital computation ? Can we control the effects of digitisation rather better ? As Shamos asked, what geometric operations are appropriate for digital computation? If we chose a different model of geometry, what benefits might accrue ? Standard PCs and workstations now provide a level of graphics rendering quality which is more than adequate for current displays. This is provided through special purpose graphics hardware which typically renders geometry reduced to polygonal arrays or meshes. In systems using traditional scan conversion, geometry is thus subjected to two stages of approximation: surfaces are approximated by planar polygon arrays, and the polygons are then rendered as pixel arrays. As the hardware becomes faster, the trend is not to render more frames per second (video and movie rates or a small multiple of these rates suffice for human perception) but to render increasingly large numbers of polygons. Display size and resolution has altered little in the last three decades and may not increase much for some time to come. It would be surprising if there was a need for orders of magnitude higher pixel counts due to the limitations of the human perception system. Simple calculations show that for some of the models currently rendered, the polygon count is so high that a single polygon on average accounts for a single pixel or even a single bit of intensity for a single pixel. We are thus reaching the stage where polygons, already defined in terms of three or more points, effectively act as single points because of their projected area in screen space. Recognising this, in recent years, a variety of techniques have been developed for point-based rendering [Westover, 1998, Laur and Hanrahan, 1991, Rusinkiewicz and Levoy, 2000, Forrest, 1979]. Surfaces can be rendered directly from points using well-chosen reconstruction filters, for example by splatting.

Thus there is a case for the common representation of geometry to be based on point arrays rather than polygon meshes. If point-based rendering is practical, then perhaps point-based modelling is a useful avenue to pursue. Recently there has been some interest in developing techniques based on collections of points (essentially polygon vertices or mesh vertices) defined by real coordinates in 3D space [Pfister et al., 2001, Schroeder et al., 2001].

Here by contrast we suggest a different approach based on integer coordinates and regular grids. Sabin, independently, has begun to explore along the same lines [Sabin, 1999]. Our starting point is the work of Corthout and Pol [Corthout and Pol, 1992]: rather than scan convert, they ask, for each pixel, whether that pixel is part of the object to be rendered, an approach called point containment. They develop a theory of discrete Bezier curves defined on an integer grid. This includes a Jordan curve theorem for regions bounded by discrete curves. The mathematical theory leads to algorithms in a formal specification language, and proofs of correctness of the algorithms with all special cases handled robustly and correctly. The necessary precision of the integer grid can readily be determined, in their case for rendering on a screen or printer. Algorithms were implemented and tested in C and in silicon as a VLSI chip which provided most of the geometric functionality of PostScript Level 1: regions defined by Bezier curves are filled and curves can be stroked by brushes defined by closed collections of non-rational and rational Bezier curves. Fabris and Forrest later extended Corthout and Pol's approach to robust anti-aliasing of Bezier curves [Fabris, 1995, Fabris and Forrest, 1997].

Whilst point containment was developed for graphical rendering, it is an equally valid approach, perhaps at rather higher resolution, to point-based modelling. For many geometric objects, efficient and robust algorithms exist for their evaluation in terms of points on a regular integer grid. We can tackle the discretisation issue at an early stage in such a system. Just as Levoy and Whitted suggest points as display primitives [Levoy and Whitted, 1985] we might investigate points as low-level modelling primitives, doing away with the tyranny of the polygon. Points can provide a common format at the lowest level for both constructed and scanned geometric data. Integer space is more amenable to analysis and proof. Discrete versions of the commonly employed geometric operations will be required (Corthout and Pol's discrete Jordan Curve containment algorithm is one example), and better control of robustness should be possible.

6. Conclusions

Many problems and uncertainties in computational geometry arise from trying to implement Euclidean geometry in a non-Euclidean space, for example floating point. A better choice of geometry may be the answer to geometric uncertainty.

References

[Braid, 1973] Braid, I. (1973). *Designing with Volumes*. PhD thesis, University of Cambridge, Computer Laboratory.

[Braid, 2001] Braid, I. (2001). Personal communication.

[Chazelle and Dobkin, 1980] Chazelle, B. and Dobkin, D. (1980). Detection is easier than computation. In *12th ACM Symposium on the Theory of Computation*, Los Angeles, California.

[Corthout and Pol, 1992] Corthout, M. and Pol, E.-J. (1992). *Point Containment and the PHAROS Chip*. PhD thesis, University of Leiden, Leiden.

[Fabris, 1995] Fabris, A. (1995). *Robust Anti-aliasing of Curves*. PhD thesis, University of East Anglia Computational Geometry Project.

[Fabris and Forrest, 1997] Fabris, A. and Forrest, A. (1997). Antialiasing of curves by discrete pre-filtering. In *Computer Graphics Proceedings, SIGGRAPH*, pages 317–326, Los Angeles, California. ACM SIGGRAPH, New York.

[Forrest, 1971] Forrest, A. (1971). Computational geometry. *Proceedings of the Royal Society of London A*, 321:187–197.

[Forrest, 1979] Forrest, A. (1979). On the rendering of surfaces. In *Computer Graphics Proceedings, Annual Conference Series, SIGGRAPH*, volume 13, pages 253–259.

[Laur and Hanrahan, 1991] Laur, D. and Hanrahan, P. (1991). Hierarchical splatting: A progressive refinement algorithm for volume rendering. In *Computer Graphics Proceedings, Annual Conference Series, SIGGRAPH*, volume 25, pages 285–288. ACM SIGGRAPH, New York.

[Levoy and Whitted, 1985] Levoy, M. and Whitted, J. (1985). The use of points as display primitives. Technical Report Technical Report TR 85-022, University of North Carolina at Chapel Hill, Department of Computer Science.

[Pfister et al., 2001] Pfister, H., Rockwood, A., Frisken, S., Perry, R., Gross, M., McMillan, L., Moreton, H., and Sweldens, W. (2001). New directions in shape representations. In *Course 33: On SIGGRAPH 2001 Course Notes CD-ROM, ACM SIGGRAPH*, Los Angeles, California.

[Rusinkiewicz and Levoy, 2000] Rusinkiewicz, S. and Levoy, M. (2000). Qsplat: A multiresolution point rendering system for large meshes. In *Computer Graphics Proceedings, Annual Conference Series, SIGGRAPH*, pages 343–352, New Orleans, Louisiana. ACM SIGGRAPH, New York.

[Sabin, 1999] Sabin, M. (1999). Explorations in 3D integer-based linear geometry. Technical Report Technical Report DAMTP/1999/NA05, University of Cambridge, Department of Applied Mathematics and Theoretical Physics.

[Schroeder et al., 2001] Schroeder, P., Sweldens, W., Curless, B., Guskov, I., and Zorin, D. (2001). Digital geometry processing. In *Course 50: On SIGGRAPH 2001 Course Notes CD-ROM, ACM SIGGRAPH*, Los Angeles, California.

[Seidel, 1998] Seidel, R. (1998). The nature and meaning of perturbations in geometric computing. *Discrete and Computational Geometry*, 19(1):1–17.

[Shamos, 1974] Shamos, M. (1974). Problems in computational geometry. Ph.D. Thesis Outline.

[Shamos, 1975] Shamos, M. (1975). Geometric complexity. In *7th SIGACT Conference*.

[Shamos, 1978] Shamos, M. (1978). *Computational Geometry*. PhD thesis, Yale University, Department of Computer Science.

[Solomon, 1985] Solomon, B. (1985). *Surface Intersections for Solid Modelling*. PhD thesis, University of Cambridge.

[Westover, 1998] Westover, L. (1998). Footprint evaluation for volume rendering. In *Computer Graphics Proceedings, Annual Conference Series, SIGGRAPH*, volume 24, pages 367–376.

Chapter 7

GEOMETRIC UNCERTAINTY IN SOLID MODELING

Pierre J. Malraison
PlanetCAD, Inc.,
Boulder, Colorado 80301, USA
pierre.malraison@planetcad.com

William A. Denker
7382 Mt. Sherman Rd.
Longmont, Colorado 80503, USA
bill_denker@yahoo.com

Abstract In solid modeling geometric uncertainty arises in several settings. For systems which use analytic solutions when possible, it is necessary to determine the type of surface so that the correct analytic solver will be used. In passing from 3-D space data to 2-D parametric data, the algorithms used are vulnerable to geometric uncertainty – examples and solutions from a project in a commercial solid modeler [ACIS, 2001] are presented. For any solid modeler, there is a fundamental uncertainty imposed by the use of exact logic for topology together with necessarily inexact logic for geometry.

Keywords: Solid modeling, data translation, healing, parametric surfaces

1. Introduction

This paper covers three types of geometric uncertainties arising in the context of geometric modeling. First, uncertainty as to the type of surface, e.g., sphere or very un-eccentric ellipsoid. Second, uncertainty about approximations that arise from the mapping between parameter space and three-space. Third, an epistemological uncertainty caused by the fact that computer algorithms involving topology demand exact answers but the underlying geometric mechanisms only operate within a given tolerance.

2. "Pure" Geometry

The classical example of geometric uncertainty in this area is the quadratic formula. In theory (or for an infinite precision computer) it provides an exact solution to finding the roots of a quadratic. In practice it becomes numerically unstable close to singular points. This section presents two other examples of this sort of phenomenon occurring in geometric modeling.

2.1 Quadrics

[Levin, 1976] describes how to draw and intersect quadric surfaces using the discriminant of the matrix for the quadric. [Hakala, et. al. 1980] discuss the geometric uncertainties arising from an actual implementation.

The technique is to represent a quadric surface as a 4×4 matrix Q. If x is a vector in homogeneous coordinates, the quadric surface is defined by $xQx^T = 0$. Levin's algorithm then looks at the matrix $P + \lambda Q$ having the simplest form to draw the curves on $P \cap Q$.

The uncertainty is: how to decide if Q is *almost* a sphere; do we treat it as a sphere or an ellipsoid? Let Q' be the upper left 3 x 3 part of Q. For a sphere, Q' is diagonal with all the diagonal entries equal to 1. If the diagonal elements are very close to one rather than exactly one, the quadric surface is an ellipsoid. If the off-diagonal elements are not exactly zero, the quadric could be anything. In [Hakala, et. al. 1980] the issue was avoided by restricting attention to the "natural" quadrics: planes, spheres, cylinders and cones.

2.2 When is a cone too flat?

One algorithm for intersecting a cone with a plane looks at the ellipses formed by intersecting a bounding box with a cone. It projects one corner of the bounding box onto the axis and uses that point to determine one of the planes for the ellipses. If the cone angle is near $90°$ the radius of those ellipses can be very large. These anomalies arise typically not by explicitly making a cone but by revolving a profile about an axis. The bounding box in this case is usually far from the vertex. If α is the cone angle and h is the distance from the vertex, the ellipse radius is $h\,tan(\alpha)$. For $h = 10^3$ and $\alpha = 89°, r = 114,571$. This is not a "large" number, but is large enough to cause numerical issues when the rest of the numbers (e.g. the size of the bounding box) are smaller.

The definition of "nearness" is arbitrary - in the software we use, the test was that the cosine of the cone angle should be less than 0.1. This translates to about $85°$. In that case we use a different algorithm to find the intersection: project all of the corners of the bounding box onto the cone axis and sort those values to find a better ellipse.

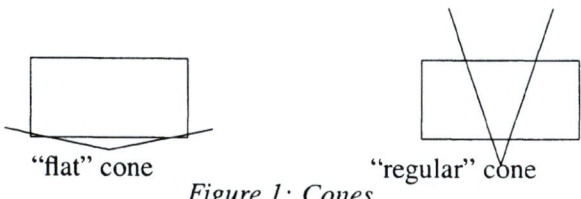

Figure 1: Cones

For very small values of the cosine of the cone angle it is better to treat the cone as a plane. The threshold used for this is quite tight: 10^{-10}. This is an arbitrary choice but has proven to be adequate in practice.

2.3 Conic Sections

There are several different ways to represent ellipses, parabolas and hyperbolas, and none is stable for all conic-section curves. The classification of a conic can be quite significant, because the geometric properties and algorithms can be very different for the three types: ellipses have a center and two axes, parabolas have a focus and a directrix, etc. The method used to classify a conic depends on the representation. For example, in the common algebraic representation $Ax^2 + Bxy + Cy^2 + Dx + Ey + F = 0$, the *discriminant* $B^2 - 4AC$ is invariant under translation and rotation. This value is negative, zero or positive for ellipses, parabolas or hyperbolas, respectively. In the corresponding cases, the *eccentricity* is less than, equal to, or greater than 1. When represented as a rational quadratic with weights w_0, w_1 and w_2, a conic is a parabola whenever the term $w_0 w_2 / w_1^2$ is exactly 1. If a conic were represented as the intersection of a 3D conic surface and a plane, the parabola case is when the cone axis is exactly parallel to the plane. In all cases, the curve is a parabola if one real number "equals" another. The value of a floating-point tolerance will depend on the representation and the application, and choosing an appropriate value can be very difficult. Wilson [Wilson, 1987] provides a more detailed discussion.

3. Surface Inversion

This section presents examples of problems caused by geometric uncertainty that arose during the development of a geometric algorithm within a commercial solid modeler (ACIS [ACIS, 2001]). Some of the problems were expected, and some were not.

The task was to invert curves on a surface: given a parametric surface $S(u,v)$ and a three-space curve $C(t)$ lying on (or very near) the surface, create the preimage $P(t)$ of the curve in the parameter space of the surface; i.e., $S(P(t)) = C(t)$, within a given tolerance. The curve $P(t)$, whose range consists of uv positions in the domain of $S(t)$, is termed a *pcurve*.

Surface inversion is central to this algorithm: given a position xyz on the surface, find a parameter position uv such that $S(u,v) = xyz$. The xyz position may not be exactly on the surface, due to noisy data, but it should be close. In the usual case, inverting a position requires finding a good uv guess if none is provided, and performing a two-dimensional Newton-Raphson iteration to set to zero the distance from the given position to the surface normal vector.

Vectors, as well as positions, must be inverted. Having inverted a curve position $C(t)$ onto the surface, so that $S(u,v)$ corresponds to $C(t)$, the derivative of uv with respect to t is required. This amounts to writing a vector (C_t) as a linear combination of two basis vectors (S_u and S_v), i.e., find u_t and v_t such that $C_t = u_t S_u + v_t S_v$. In the usual case, inverting a vector requires solving a two-by-two linear system. In the current algorithm, the parameter-space derivative uv_t is used to find the coefficients of the two-dimensional B-spline for the pcurve.

Each Newton step for inverting a position is actually the same problem as inverting a vector, but there are differences in the requirements of the two operations that necessitate differences in the code. The main difference occurs when the inversion operation – writing a three-dimensional vector as a linear combination of two basis vectors – results in a very large step in uv space. In a Newton iteration, this uv step is used to change a uv guess value, and large steps generally represent a problem in the progress of the algorithm (which are to be expected near singularities, for instance). But when converting a three-dimensional derivative vector into the derivative vector of a corresponding two-dimensional parameter-space curve, a large uv vector is perfectly acceptable, as long as it is accurate. For this reason, system routines were sufficient for position inversions, but a special routine had to be written to invert a vector.

3.1 Expected Complications

3.1.1 Poles.
A surface is singular when the surface parametric normal vanishes, i.e., $S_u \times S_v$ is the zero vector. The most common reason for this is that one of the surface derivatives goes to zero. Singularities include the poles of spheres, apices of cones, and the degenerate edge of a three-sided patch represented as a four-sided tensor-product spline.

As an example, consider a sphere, where u is the longitude, $0 \leq u < 2\pi$, and v is the latitude, $-\pi/2 \leq v \leq \pi/2$. Then at the north pole, S_u is the degenerate derivative, and S_v runs into the pole. When inverting an xyz point corresponding to the pole, any parameter position $(u, \pi/2)$ will map to that position, and the system inversion routines will indeed return any value. For our purposes, however, the value of u is significant: the pcurve should be well behaved in parameter space. For example, an incorrect u value can cause what should be an isoparametric curve to take drastic turns near the pole. Analogously, the vector-inversion method must return the correct uv_t, so that the derivative of the pcurve is correct in parameter space.

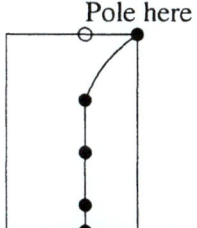

Figure 2: *Parametric pole*

The basic solution to this is quite simple, and almost always works: just make u the same as that of an adjacent point. The corresponding solution for vector inversion is to set the degenerate component of the parameter derivative to zero. This makes the pcurve go straight into the pole along an isoparameter line, which in real-world models is almost always the case.

3.1.2   Seams.   Periodic surfaces add another complication. The sphere again provides a good example: at any point along the prime meridian, $S(0, v) = S(2\pi, v)$, and an inversion algorithm could return either 0 or 2π as the u value.

There are two cases here: the curve runs either along the seam, or across it. If the curve is the seam curve, our algorithm deals with the ambiguity by always returning the low edge of the period; at this level, there is no information to decide otherwise. If the curve runs across the seam, then we may assume that a point on the seam is either the start or end of the curve. In this case, we check whether its direction corresponds to increasing or decreasing u on the surface.

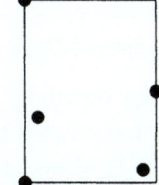

Figure 3: *Curve running along a seam*

3.1.3 Incorrect Convergence. It is also possible for inversion routines to converge to an incorrect solution. Because the position to be inverted does not necessarily lie on the surface, the convergence criterion is actually that xyz is on the surface normal vector, within tolerance. Therefore, the inversion algorithms can converge to a local distance maximum instead of a minimum. This was observed, for example, on a large, thin torus. A special method was written to determine whether a uv parameter position is the proper inversion of an xyz position. It checks both whether the uv is a solution, and whether it appears to be the correct solution.

3.2 Unexpected Complications

Even the expected complications such as periodicity and singularities became problematic when dealing with real-world data.

3.2.1 Behavior Near Poles. In addition to dealing with singularities, points that are very close to true singularities also need special handling. If one derivative is extremely small compared to the other, then Newton iteration steps can be meaningless. As an example, consider inverting a position very near the pole of a sphere. The standard two-dimensional Newton step can, because of numerical noise, give large steps in u. These large steps have almost no effect on the surface position, but they can prevent the algorithm from settling into a solution. This behavior arises when the three-space step is closely aligned with the non-degenerate surface derivative. In that case, a tiny component in the direction of the degenerate component, which could easily be nothing more than numerical noise, can cause a macroscopic, and meaningless, change in parameter value.

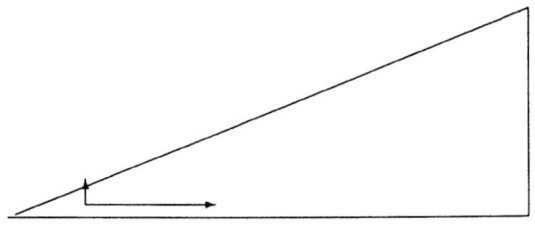

Figure 4: Parametric Degeneracy

3.2.2 Detecting Poles. Even just deciding whether a uv parameter position corresponds to a singular point has caused problems. There are two reasons for this. First, ACIS [ACIS, 2001] allows surface singularities only on domain boundaries. Because of this, system methods first check whether the parameter position corresponds to a surface parametric boundary, and if not, report that the point is nonsingular. But imported data has been known to contain internal singularities, for example where the surface passes through a pole and emerges, flipped, on the other side.

Another reason why it can be difficult to determine whether a point is singular is the age-old question, how small is a "zero" vector? Legitimate derivatives have been encountered with magnitudes as small as 10^{-7} and as large as 10^7. This would indicate that tests for degeneracy should be relative, not absolute. The basic criterion that is appropriate for this application is the ratio of the magnitudes of the derivatives: if S_u is, say, 10,000 times as big as S_v, then this parameter position is presumably unstable and special care should be taken. Unfortunately, even that is not enough in practice: counterexamples arise in which absolute tests are required in addition to the relative tests. In one case, a customer would routinely scale parts down by 0.0001, which gave rise to ratios of terms such as $4.235 \times 10^{-14} / - 4.235 \times 10^{-14}$, necessitating a relative test. But in another case, subtractive cancellation resulted in ratios such as $1.047 \times 10^{-24} / 3.289 \times 10^{-21}$. These were both were meaningless numbers, but the ratio happened to be reasonable. For these reasons, more general methods for determining singularity had to be written.

3.2.3 Pcurve Position and Direction at Poles.

The new pcurve algorithm was occasionally observed to create pcurves with very many small segments, and not particularly good accuracy, as they came into poles. The reason turned out to be that the curves were "curving" into the pole, in parameter space. As described under expected complications, this is very rare in practice. New methods were required to find the correct value – and derivative – of the degenerate parameter at the pole.

In any event, a special method was written to decide whether a parameter-space derivative is accurate. It simply checks the result in three dimensions: it compares the given three-space vector with the composition $u_t S_u + v_t S_v$. As soon as the check routine is satisfied, the result is returned. If it is not able to calculate any result that will satisfy the check routine, this fact is communicated to the calling routines, which are then able to use other information to find an acceptable value to be used in the pcurve spline.

3.2.4 Seams.

As with poles, what exactly does it mean to be "on" a seam? Periodicity checks might invoke system methods to determine whether a value is within an interval. For a parameter value t in an interval $[a, b]$, system routines will return $a \leq t \leq b$, with the interval perhaps even expanded by some epsilon. For our application however, being within the base period of a periodic entity generally means $a \leq t < b$. For reasons such as these, local versions of some system utility algorithms had to be written.

One numerical problem arose from a very tiny curve. The curve crossed a seam in a generally perpendicular direction, and was longer than the system tolerance, but so short that every point on it was within tolerance of the seam. This somewhat pathological example necessitated a change in the algorithm for determining whether a curve represents a surface seam.

3.2.5 Bad Guesses. Newton-type iterations are famous for having extremely fast convergence, if the guess is good, but also for the possibility of pathologically bad behavior if the guess is not close enough. The caller may or may not provide a guess, and if it does, the guess could be too far away. This could be because the caller simply uses a previous value as a guess, which would usually represent a good guess, but not always. For example, if the curve corresponds to an isoparametric longitude line on a sphere, steps can be very big.

A special method was written to check whether or not to use a given guess. (If no guess is given, the system inversion routines will find one themselves. That procedure is reliable, but slower, because it has less information.) It first checks the proximity of the points: if they are very close already, it returns TRUE. It then checks a trial Newton step, and if that is too large, it returns FALSE. After that, it looks at curvatures to see whether a Newton step could converge to an incorrect solution, such as a relative maximum.

3.2.6 Incorrect Convergence. Convergence to an incorrect solution could also be expected, but the exact nature of the problems and the manner of dealing with them were not obvious during the design phase, only when encountered.

Two factors conspired to make it difficult to determine whether a given uv is the correct solution: the input curves can be a substantial distance from their surfaces, and surfaces can be highly curved (such as the large, thin torus). These two possibilities make it impossible to determine a single tolerance value for the distance between $S(u, v)$ and xyz: the distance between the input curve and its surface can actually be greater than the diameter of a surface.

The convergence-checking method must take this into account. It actually uses two different tolerance values: one to check whether the given uv is a solution at all, and one to check whether it is the correct solution. For two tolerance values called good_tol and bad_tol, the algorithm is:

- IF $S(u, v)$ and xyz are themselves within good_tol, return TRUE;

- IF xyz is not on surface normal to within good_tol, return FALSE;
  // This is a solution; check whether it is the correct one.
  // Check the surface curvature:

- IF xyz is at or beyond the center of curvature of the surface, return FALSE; // wrong convergence;

- IF the points are farther apart than bad_tol, return FALSE; // wrong convergence;

The two tolerance values must be determined by the caller, and could depend on the geometric situation.

4. Topology and Geometry

This is the central quandary of building boundary representations on a computer: *Topology is from Mars, Geometry is from Venus*. Topological relationships are logical – two faces either are or are not adjacent. Geometric relationships are numerical – point a is the same as point b if the distance between them is within some tolerance. This problem can be managed within the context of a single modeler but becomes more pronounced when dealing with data created externally. For example, translating a CATIA model which was created with a tolerance of 10^{-3} into ACIS [ACIS, 2001] with a tolerance of 10^{-6}.

There are two approaches to solving this problem: healing and tolerant topology. Several companies have addressed one or both of these approaches, typically with proprietary algorithms. We present here an overview.

4.1 Healing

Because of tolerance differences, edges or faces that are supposed to be adjacent appear not to be. Most healing software breaks the process into two pieces: an analysis phase which finds the parts of a solid that need to be healed, and a healing phase which does the healing.

Healing attempts to correct a wide range of problems. Here is a list selected from the Theorem Solutions web site [Theorem, 2001]:

- Remove a Sliver Face, with a maximum width of less than a defined value.

- Heal a whole body - i.e. automatically modify edge curves and surfaces to close gaps between the edges / surfaces and vertices / curves.

- Remove a Small Edge if the length is less than a defined value.

- Remove a spike. This merges two adjacent edges of a face where the angle between the two edges and the maximum width between the two edges is small (e.g. removal of spike).

- Split edges and redefine loops etc. at a face waist position i.e. if two edges approach and are pinched other than at a mutual vertex position within a given tolerance, split the edges if necessary at that point.

- Sew a group of unconnected faces with a given sewing tolerance.

4.2 Tolerant topology

To continue the analogy of our quotation, this approach moves topology to Venus. In other words the notion of tolerant topology is introduced, typically at the vertex and edge level.

The basic approach is to add a tolerance to the topological object:

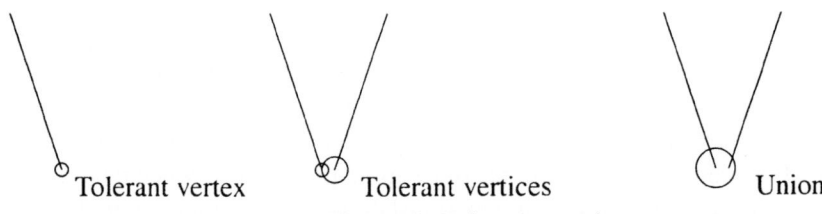

Tolerant vertex   Tolerant vertices   Union

Figure 5: Tolerant topology

When two tolerant objects interact they inherit a larger tolerance. From a theoretical point of view, this would seem to lead inevitably to all objects being reduced to a single tolerant vertex. In practice, tolerant vertices tend to be isolated to fairly local parts of a solid model, for example when doing variable radius blends with small radii.

In addition to tolerant vertices a tolerant modeler needs to deal with tolerant edges. This means making explicit the fact that when two faces intersect there are five distinct curves related to the intersection:

- the 3-space intersection curve,
- the two parameter space curves created by projecting the intersection curve onto the underlying surfaces of the face,
- the two 3-space curves obtained by re-evaluating those parameter space curves.

These curves are always "the same" within a tolerance, but in a tolerant modeler algorithms must keep track of all five and determine which one to use. For example, blending an edge typically uses the intersection curve since the blend will usually generate new edges on the two faces.

5. Acknowledgements

We would like to thank the organizers of the *Uncertainty in Geometric Computations* Workshop for accepting our paper, PlanetCAD for support in the preparation and presentation of the talk, and the referees for helpful comments.

References

Spatial Corporation, a Dassault Systemes, S.A. company *ACIS 3D Geometric Modeler (ACIS)*,
http://www.spatial.com/products/Toolkit/toolkit.htm.

D.G. Hakala, R. C. Hillyard, P. Malraison, B. E. Nourse, *Natural quadrics in mechanical design,* Proc - AUTOFACT West, Vol. 1, CAD/CAM 8, Anaheim, Calif, USA, November 17-20, 1980 Publ by SME, Dearborn, Mich, USA, 1980, 363–378.

J. Levin, *A parametric algorithm for drawing pictures of solid objects composed of quadric surfaces,* Communications of the ACM, Vol. 19, No. 10, 1976, 555–563.

Theorem Solutions, Ltd. *CADhealer information,*
http://www.theorem.co.uk/docs/cadheinfo.htm.

P. Wilson, *Conic Representations for shape descriptions,* IEEE Computer Graphics and Applications, Vol. 7, No. 4, 1987, 23–30.

Chapter 8

RELIABLE GEOMETRIC COMPUTATIONS WITH ALGEBRAIC PRIMITIVES AND PREDICATES

Mark Foskey, Dinesh Manocha
University of North Carolina at Chapel Hill, North Carolina, USA
{dm,foskey}@cs.unc.edu

Tim Culver
Think3, Sudbury, MA, USA
culver@acm.org

John Keyser
Texas A&M University, College Station, Texas, USA
keyser@cs.tamu.edu

Shankar Krishnan
AT&T Labs—Research, Florham Park, New Jersey, USA
krishnas@research.att.com

1. Introduction

The problem of accurate and robust implementation of geometric algorithms has received considerable attention for more than a decade. Despite much progress in computational geometry and geometric modeling, practical implementations of geometric algorithms are prone to error. Much of the difficulty arises from the fact that reasoning about geometry most naturally occurs in the domain of the real numbers, which can only be represented approximately on a digital computer. Many times, the correctness of geometric algorithms depends on correctly evaluating the signs of arithmetic expressions, and errors due to rounding or imprecise inputs can lead to grossly incorrect results or failure to run to completion.

The proposed solutions to this problem can be classified into *inexact* and *exact* approaches [28]. The former approach accepts the inaccuracy of the machine representation, and attempts to modify the algorithms, given that constraint, so that they reliably produce acceptable output. The notion of acceptability is dependent on the application. Algorithms developed in this way have been shown to work in specific cases. On the other hand, *exact geometric computation* (EGC) requires that every predicate evaluation be correct [34]. The exact computation paradigm eliminates numerical error in geometric computations entirely. Unfortunately, exact implementations are often far too slow, especially when we are dealing with nonlinear primitives. Karasick et al. [21] noted that naive implementations can take several orders of magnitude longer than an equivalent floating-point implementation, an observation that is consistent with our experience. The goal has been to find techniques that reduce the performance penalty to an acceptable level.

As work on exact geometric computation has proceeded, it has become clear that the performance problems can be greatly alleviated. One area that has received less attention is the issue of reliability when dealing with nonlinear algebraic or curved primitives. This area provides numerous interesting challenges for EGC and we address some of them in this paper.

Exact Computation as a Practical Approach. There is no question that EGC is slower than computation relying solely on machine precision arithmetic. The question is whether the slowdown is worth the gain in precision. Indeed, in many scientific or engineering applications the input data is inexact, and the question arises whether an exact result is even meaningful. But the main reason for using EGC is not exactness in itself, but rather *reliability*. A common cause of program failure is that rounding errors lead to inconsistent combinatorial decisions, e.g. about where a point lies with regard to a surface. By making a single interpretation of the data and performing calculations that are consistent with that interpretation, we can avoid this source of failure. Solving the problems of accuracy and consistency is the first step towards a general solution to the robustness problem, which also involves handling degeneracies and special cases.

Organization. The rest of this paper is organized as follows. In Section 8.2 we briefly review the relevant literature. In Section 8.3 we present some underlying geometric problems involving curved primitives. We discuss our general approach to EGC for curved primitives in Section 8.4, including methods we use to achieve reasonable speeds for these computations. In Section 8.5 we present some results for boundary evaluation and Voronoi computations, and we conclude in Section 8.6.

2. Literature Survey

The issue of robust and accurate computations in geometric applications has been addressed in numerous places, with surveys by Hoffmann [18] and Fortune [15] giving an indication of the variety of work. Yap [33] described the concept of exact geometric computation, and Li and Yap [27] have presented a more recent survey. Some of the earlier inexact approaches were based on geometric tolerances [31] and interval arithmetic [9].

One of the key ideas in accelerating exact computation is the use of *floating point filters*, in which predicates are first evaluated using fast floating point methods and then tested for reliability, by analyzing the size of the possible floating point error. If a predicate is unreliable, exact computation is performed. A good example of this method is the work of Fortune and van Wyk [16]. As a preprocess, they perform an analysis of the calculations that will be needed by a geometric algorithm, so that the accuracy of these computations can be checked quickly at run time.

A number of researchers have used exact computation for boundary evaluation. Benouamer, Michelucci, and Peroche [2] implement a solid modeler using a filtered approach that differs from that of Fortune and van Wyk. Benouamer et al. express each sequence of calculations as an *expression dag*, that is, a directed acyclic graph with operations at internal nodes and constants at the leaves. Calculations are initially performed using interval arithmetic, and if the result is not sufficiently precise, then exact rational arithmetic is used.

Fortune [14] also used exact arithmetic to implement a polyhedral solid modeler. That work sets an upper limit on the bit-length of accepted input, so that all geometric predicates can be evaluated using arithmetic at some fixed precision.

Yap and Dube [34] introduced a general approach they call "precision-driven computation." Like Benouamer et al., they also use expression dags, but as a tool to determine in advance the amount of precision needed (that is, the number of digits in a floating point representation). Precision-driven computation is noteworthy because it is fundamentally distinct from filtered approaches. However, it is only applicable for closed-form calculations.

Boundary evaluation in solid modeling has been a well studied research topic in the area of polyhedral models. In addition to the results on the subject mentioned above, we note the work of Hoffmann [19] and Requicha and Voelcker [30]. Some algorithms have also been proposed for quadrics or higher degree algebraic primitives. Casale et al. [4] use trimmed parametric surfaces to generate boundary representations of sculptured solids. Their algorithm uses subdivision methods to evaluate surface intersections, and represents the trimming boundary with piecewise linear segments. Krishnan and Manocha presented algorithms and a system called BOOLE based on the algebraic formulation of the problem [26]. BOOLE is based on lower dimensional algorithms for computing the intersections of parametric surfaces and uses a combination of symbolic and numeric algorithms [25]. It uses 64-bit IEEE floating point arithmetic.

One area where reliability is particularly challenging and has received relatively little study is computation of the medial axis of a polyhedron. There have been a number of approximate approaches, however. Vleugels and Overmars [32] and Etzion and Rappoport [13] both use recursive subdivision of space to create an arbitrarily close approximation, while Hoff et al. [20] compute an approximation at a fixed resolution using graphics hardware. Other authors do not rely on an approximation. Of these, Milenkovic [29] was the first to propose an algorithm for computing the medial axis as a 3D geometric object by tracing the seams between the curved faces of the structure.

3. Nonlinear Geometric Problems

In this section we discuss a number of problems involving curved geometric primitives that arise in geometric applications.

3.0.1 Polynomial Root Isolation.
The isolation of complex roots of polynomials is in a sense a geometric problem in the complex plane. Also, the other problems we discuss will depend in an essential way on localizing the real roots of polynomials.

3.0.2 Curve Arrangements.
Given a number of algebraic curves in a bounded region of the plane, the goal in the curve arrangement problem is to compute the connected subregions that have no curve passing through them. Each subregion, called a *face*, is defined by piecewise algebraic curves that enclose its boundary. For each of the polynomials defining a face, all points in the face will have the same sign with regard to that polynomial. The output of the curve arrangement algorithm is the explicit topological description of each cell.

3.0.3 Boundary Evaluation.
In computational solid geometry (CSG) [19], objects are constructed from solid primitives by the boolean operations of union, intersection, and set difference. A complicated object will be represented as a tree, with geometric primitives at the leaves, and boolean operations at the internal nodes. The *boundary evaluation problem* is the problem of taking a CSG model and constructing from it a representation of its boundary as a set of possibly curved two dimensional surface primitives with adjacency information.

3.0.4   The Medial Axis Transformation.   The medial axis of a polyhedron is the locus of points that are the centers of spheres contained in the polyhedron and touching the boundary at two or more points. In 3 dimensions, the medial axis is made up of portions of quadric surfaces intersecting along curves. All of the known practical algorithms for computing the exact medial axis—explicitly constructing all of its surfaces and curves—rely on tracing these curves, starting at the vertices of the polyhedron [6]. Combinatorial errors at early stages can cause incorrect curves to be generated, typically resulting in a program failure. The probability for such a failure increases rapidly with the complexity of the polyhedron.

4.   Exact Geometric Computation for Curves and Surfaces

The fundamental challenge of EGC in the curvilinear domain is that the coordinates of points determined by intersecting polynomial surfaces will typically be irrational and thus not representable exactly by a rational package. We represent such a point by retaining the polynomials defining it, along with an axis-aligned box with rational coordinates known to contain only that intersection. There are mechanisms to shrink the box when necessary, to isolate the roots and make comparisons between the points. As a result, the algorithm uses minimal precision to accurately perform the tests. This technique is an example of the distinction between exact arithmetic and exact computation—the only explicit representation of a point is inexact, but all comparisons are made exactly. The library MAPC [22] has been designed to embody these representations for points and curves in two dimensions.

Methods used to isolate points in 2D and 3D are detailed in [23] and [6] respectively. We will briefly discuss them here to indicate the kind of calculations needed. The fundamental tool we use for reliably localizing algebraic points is the *Sturm sequence*. For a univariate polynomial f, the Sturm sequence of f is the polynomial remainder sequence of f and f', with the signs changed according to a simple convention. For any given real number x, the number of *sign permanencies* $\text{PERM}(f, x)$ is the number of times the sign remains the same when successive polynomials in the Sturm sequence are evaluated at x. For two real numbers $x < y$, $\text{PERM}(f, y) - \text{PERM}(f, x)$ is the number of real roots of f between x and y.

There are generalizations of the Sturm sequence concept for sets of polynomials in two and three dimensions, but they are much slower. In the computation of the medial axis, there are times when a point must be localized in 3D using the trivariate Sturm sequence [6]. However, in most cases it is possible to localize a 2D point using only univariate Sturm sequences [23]. The idea is to use an alternate method to find candidate boxes that may contain one or more points, and then to reduce the problem to a sequence of root determinations on the boundaries of these boxes.

4.1 Improving Efficiency

To ensure accuracy, Sturm sequence calculations are usually done with exact arithmetic. Since bit lengths arising in a Sturm sequence calculation can be exponential in the degree of the polynomial, these calculations can be quite slow. To improve running times, we have made extensive use of floating point filters. Unfortunately, there is a large gap between the 53 mantissa bits of a machine double on current hardware and the hundreds of bits that can arise in a Sturm sequence calculation. It is useful to have an arithmetic that is faster than rational arithmetic but that can be flexible in the amount of precision it allows. Aberth and Schaefer [1] have proposed a solution to this problem in what they call *range arithmetic*, which combines conservative interval arithmetic with variable precision floating point computation. Each number in their arithmetic is represented by a single floating point number of arbitrary length, with an associated single-precision radius R. The radius R indicates the width of the associated interval, on the scale of the least significant machine word in the representation of the floating point number.

To guarantee the precision of their results, Aberth and Schaefer perform a calculation at a specified initial precision, keeping track of the loss of accuracy. If the accuracy of the result is insufficient, they increase the precision of their representation and recompute the results with increased precision. We find that this approach of *iterative revision* is useful for many geometric problems, where precision-driven computation using expression dags may not be effective.

5. Some Results

In this section, we give examples of how the techniques we have described can be applied to a number of problems.

5.1 Polynomial Root Isolation

To isolate complex roots of polynomials, we have used Aberth and Schaefer's Range library [1] to apply a variant, described in [24], of the Durand-Kerner algorithm [10, 8] to a number of polynomials. A similar algorithm has also been proposed by Bini [3]. We tested this algorithm on a benchmark set of polynomials from the PoSSo project, available at the following site:

    http://www-sop.inria.fr/saga/POL

Each class of polynomials is known to be troublesome for many root finding algorithms. We compared our results with two other packages, Maple and MuPAD. The results are given in Table 8.1. Maple and MuPAD allow the user to specify the number of digits retained throughout a calculation, but not the (smaller) number of digits that will be reliable in the output. For the purposes of comparison, we specified that calculations be performed with the same number of digits as the maximum used in our calculations with the Range library. Because the Range library performs later calculations with fewer digits than the maximum required, it has a speed advantage when intermediate calculations must be performed at a much higher precision than needed in the final result. The particular root-isolation method used also contributes to the speedup. There is, however, an overhead in maintaining the error interval, which becomes more apparent when less precision is needed. These results were previously reported in [24].

We used the following polynomials:

- **Poly1:** $\sum_{i=0}^{n} \frac{x^i}{i!}, n = 50$.
- **Poly2:** $(x - 3c^2)^2 + icx^7, 0 < c \ll 1, c = 10^{-20}$.
- **Poly3:** $(c^2 x^2 - 3)^2 + c^2 x^9, c = 10^{20}$.
- **Poly4:** $x^{20} + cx^{14} + x^5 + 1, c = 10^{12}$.
- **Poly5:** $\prod_{i=1}^{n} (x - i), n = 40$.
- **Poly6:** $(0.01 x^{10} + (x - 10)^2) \prod_{i=1}^{20} (x - i)$.
- **Poly7:** $\prod_{i=1}^{20} (x - i)(x - 20)^2$.
- **Poly8:** $x^{14} + 2cx^{11} + c^2 x^8 + 4x^7 - 4cx^4 + 4, c = 10^{24}$.
- **Poly9:** $x^n - a, n = 50, a = 1$.

5.2 Determinant Sign

The problem of efficiently computing the sign of the determinant of large rational matrices has not been extensively studied, but there are situations where it can be useful [5]. One straightforward approach, using the Range library, is simply to use Gaussian elimination with partial pivoting, repeating the process with progressively higher precision until the sign can be determined reliably. Figure 8.1 indicates that this method often performs much better than exact methods that use modular arithmetic, when applied to nonsingular matrices. For general matrices, Culver et al. [7] proposed a filter for computing determinant signs exactly. The singular value decomposition, computed at machine precision, can be used to indicate whether the matrix is likely to be singular,

Case	Root Precision	MuPAD	Maple	Range
Poly1	10	78.77	4.79	13.17
Poly2	120	5.15	41.54	13.06
Poly3	80	3.32	8.539	6.32
Poly4	30	4.569	0.792	0.82
Poly5	30	32.21	36.31	5.67
Poly6	30	14.01	34.31	0.65
Poly7	30	6.517	13.53	1.23
Poly8	30	12.21	18.717	5.26
Poly9	30	71.215	1.26	1.44

Table 8.1. Univariate root finding algorithm applied to nine polynomials from the PoSSo benchmark suite. The second column indicates the number of required significant digits, specified in advance. The last three columns indicate running times taken by Maple, MuPAD and our algorithm. Times are in seconds and measured on an SGI Origin 400 MHz R12000 processor running Irix6.5.

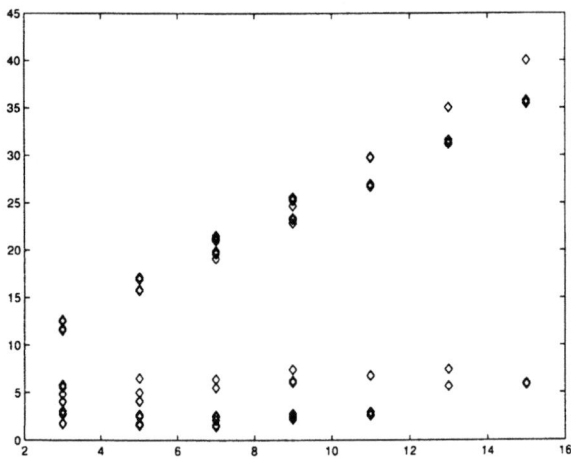

Figure 8.1. **Determinant sign speedup.** The speedup factor for the Range library, for various matrices. The horizontal axis gives the order of the matrix, while the vertical axis gives the speedup factor in comparison to an exact modular arithmetic algorithm. The set of matrices includes randomly generated matrices and others arising in geometric problems. The matrices are all nonsingular.

what the likely sign of the determinant is, and whether the matrix is well-conditioned enough to ensure that the sign estimate is correct. If the sign estimate is correct, it is used. If the matrix is ill-conditioned and likely to be singular, modular arithmetic is used. Otherwise, Gaussian elimination using the Range library is used.

Reliable Geometric Computations 99

Case	1	2	3	4
Number of Curves	3	3	6	7
Coefficient Bit size	25	22	25	62
Number of Faces	9	11	31	63
Total time without Range	5.2	9.0	62.8	122.8
Total time with Range	1.8	4.0	11.0	9.3

Table 8.2. Arrangement of planar algebraic curves. The application finds all subregions, the segments of curves bounding each subregion, and the connectivity between subregions. The table shows the maximum bit length needed to express the coefficients of the curves, the number of faces generated by the arrangement, the time taken using the original (exact rational based) code, and the time taken using Range. The curves have maximum degree 4. Times are in seconds on a 300 MHz R12000 MIPS processor.

5.3 MAPC

The MAPC library [22], mentioned in Section 8.4, provides tools for exact manipulation of algebraic points and curves. It uses exact arithmetic based on LiDIA [17] to accurately compute the required Sturm sequences. For high degree polynomials, exact computation of Sturm sequences can be prohibitively slow because of the very large growth in the bit lengths of the coefficients. Some boundary evaluation computations on algebraic primitives can require Sturm sequence computation for polynomials of degree greater than 80.

MAPC uses a number of techniques to deal with this problem. Of key importance is reducing the number of Sturm sequence computations that must be performed. For instance, floating point methods can be used to estimate the locations of polynomial roots, which are then confirmed by a Sturm sequence computation. This is much more efficient than repeated bisection using Sturm sequences.

When a Sturm sequence computation is necessary, Aberth and Schaefer's Range library [1] is used. As with the sign of determinant calculations discussed above, the process of generating the Sturm sequence and computing polynomial signs is performed at successively increasing precision until all the signs can be evaluated exactly. Once a fixed precision is reached without success (e.g. 40 decimal digits), exact rationals are used instead.

We have used MAPC as a tool in approaching each of the following three geometric problems. We give performance information for each one, with attention to the benefits of arbitrary-precision error bounded arithmetic. Again, there is a speed advantage that grows with the complexity of the problem, but a small overhead that becomes apparent with simpler calculations.

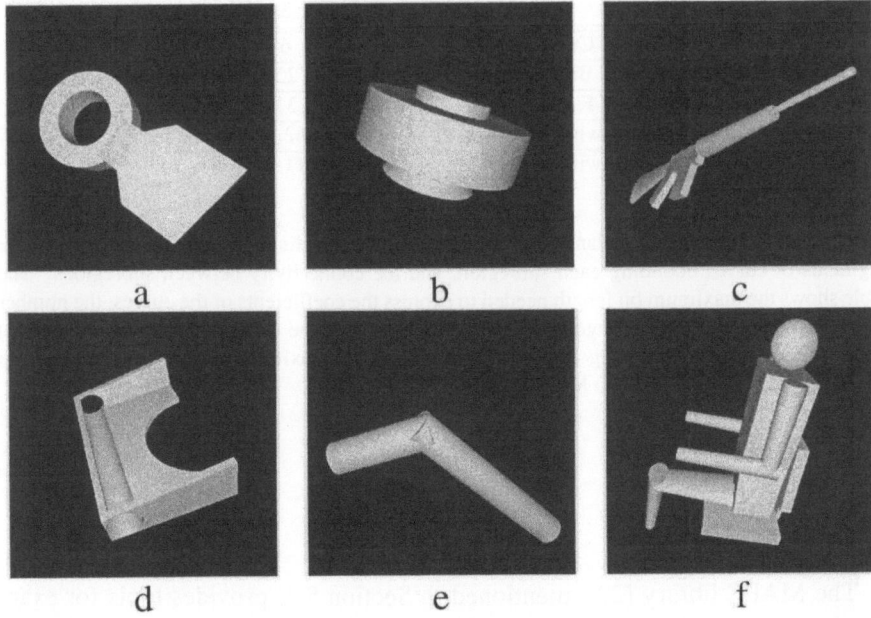

Figure 8.2. **Boundary computations.** Boundary representations of six selected portions of the Bradley Fighting Vehicle, computed by ESOLID. Computations were done exactly, and then output as trimmed NURBS patches for rendering. Computation times ranged from about 10 seconds to 633 seconds; further performance details are given in Table 8.3.

5.3.1 Curve Arrangements. The computation of curve arrangements is a useful test case for the MAPC library. In Table 8.2 we indicate performance results for some example arrangement computations.

5.3.2 Boundary Generation. A motivating application for the MAPC library is the boundary evaluation of (low degree) algebraic solids. MAPC is a core library in the ESOLID system [23] performing such boundary evaluations. In general, computing the boundary representation for a CSG model leads to problems of accuracy and robustness. These problems are exacerbated when the underlying primitives have curved boundaries. To alleviate the reliability problems, the ESOLID system performs all geometric tests exactly, using layered filters to make the exact computation more efficient.

We have tested ESOLID on portions of a real-world model, the Bradley Fighting Vehicle provided courtesy of the Army Research Laboratory. Some example output B-reps are shown in Figure 8.2, with comparative timings given in Table 8.3.

Reliable Geometric Computations

Example Number	without Range		with Range	
	Total Time	Sturm Time	Total Time	Sturm Time
a	10.23	0.51	10.95	1.62
b	12.57	0.24	12.69	1.44
c	633.42	597.33	42.99	6.93
d	63.15	8.34	61.26	6.36
e	250.74	190.62	73.86	15.36
f	26.37	1.29	28.14	3.63

Table 8.3. Timings for the examples from Figure 8.2, with and without the incorporation of the Range library. Range is used to improve the efficiency of Sturm sequence calculations. The total time and the time spent in Sturm computations is shown.

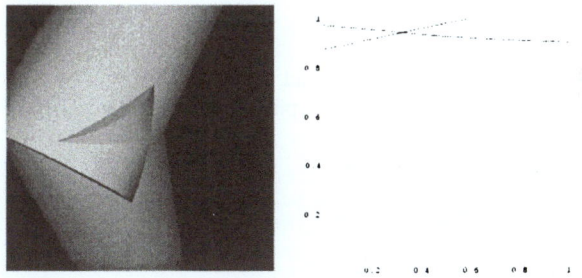

Figure 8.3. An example of a case that can be difficult for machine-precision methods. The plot on the right shows a portion of the intersection curve of the two cylinders, in the parametric domain of one of the cylinders. The curve is very nearly singular, but in fact has two distinct components.

Figure 8.3 gives an example of a calculation that can be hard for machine-precision methods. The intersection curve between the two cylinders is not self-crossing, although it appears so at the scale shown. This distinction can be hard for non-exact methods to resolve.

5.3.3 Medial Axis. Computing the medial axis of a polyhedron is a challenging problem because it inherently requires analysis of intersecting curved surfaces. Culver et al. [6] have implemented an exact algorithm for medial axis evaluation that relies both on the MAPC library and on the sign of the determinant filter described in Section 8.5.2, both of which incorporate the Range library. The program has been used to compute the medial axis of complicated polyhedra with as many as 250 faces [5]. Examples of the output are given in Figures 8.4 and 8.5. In both examples, seams are depicted as straight lines.

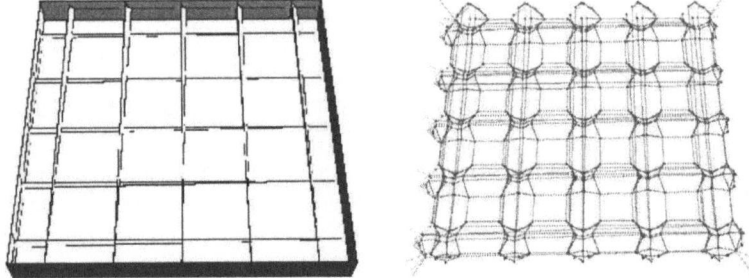

Figure 8.4. The "iron maiden pizza box" and a schematic of its medial axis. The top and bottom of the box are removed to show the spikes inside. The model has 56 faces, and the computation took 23 minutes.

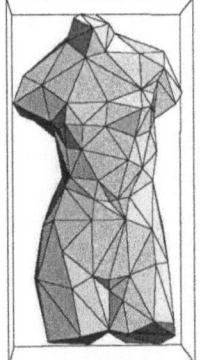 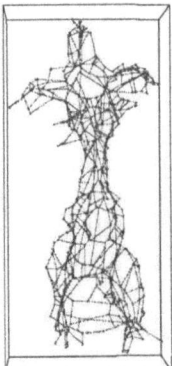

Figure 8.5. The Venus de Milo and a schematic of its medial axis. Seams touching vertices of the polyhedron are omitted for clarity. The polyhedron in this example has 250 faces, and took 5.6 hours to compute.

6. Conclusions and Future Work

We have found that exact geometric computation can be a practical tool to alleviate reliability problems in geometric computations with algebraic curves and surfaces. Some of the initial results related to root finding, curve arrangements, boundary and Voronoi computations are promising and there are many areas for future research.

One important problem is that of dealing with *degeneracies*, such as the intersection of a line with a polygon only at one vertex, or along an edge. Degeneracies can be a source of non-robustness on the one hand, or of serious implementation difficulties on the other. For simplicity, algorithms often assume that primitives are arranged so that there are no degeneracies (i.e., they are in *general position*). In practice, however, primitives are often not in general position, causing implementations of the algorithms to fail. Recasting an algorithm to handle degeneracies tends to result in a situation in which most of the code is to handle special cases. Edelsbrunner and Mucke have presented a nice overview of the problem [11].

A number of authors have proposed the idea of *symbolic perturbation* to solve these problems [11, 12]. Unfortunately, these algorithms depend on the existence of a decision tree in which each decision rests on the evaluation of an expression that is some known polynomial in the input values. In the nonlinear problems we have discussed, the calculation cannot be formulated in this way. A general approach to the problem of degeneracies, perhaps in the spirit of symbolic perturbation, would be a significant contribution.

Finally, another goal is to perform reliable computations with higher degree primitives (e.g., bicubic rational parametric patches that are widely used in geometric modeling). Currently, we have handled primitives of algebraic degree four for boundary evaluation and the medial axis computation results in quadric surfaces.

7. Acknowledgments

We would like to thank the BRL-CAD group at the Army Research Lab for the use of the Bradley Fighting Vehicle model. This work was supported in part by an ARO Contract DAAD19-99-1-0162, NSF award 9876914, DOE ASCI grant and ONR Young Investigator Award.

References

[1] Oliver Aberth and Mark J. Schaefer. Precise computation using range arithmetic, via C++. *ACM Transaction on Mathematical Software*, 18(4):481–491, 1992.

[2] M.O. Benouamer, D. Michelucci, and B. Peroche. Error-free boundary evaluation based on a lazy rational arithmetic: A detailed implementation. *Computer Aided Design*, 26(6):403–416, June 1994.

[3] Dario Bini. Numerical computation of polynomial zeros by means of Aberth's method. *Numerical Mathematics*, 13:179–200, 1996.

[4] M. S. Casale and J. E. Bobrow. A set operation algorithm for sculptured solids modeled with trimmed patches. *Computer Aided Geometric Design*, 6:235–247, 1989.

[5] Tim Culver. *Computing the Medial Axis of a Polyhedron Reliably and Efficiently*. Ph.D. thesis, University of North Carolina at Chapel Hill, Chapel Hill, North Carolina, USA, 2000.

[6] Tim Culver, John Keyser, and Dinesh Manocha. Accurate computation of the medial axis of a polyhedron. In *Proc. Symposium on Solid Modeling and Applications*, pages 179–190, 1999.

[7] Tim Culver, John Keyser, Dinesh Manocha, and Shankar Krishnan. A hybrid approach for evaluating signs of moderately sized matrices. Technical Report TR00-020, Department of Computer Science, University of North Carolina, 2000.

[8] K. Dochev. *Physical and Mathematical Journal of the Bulgarian Academy of Sciences*, 5:136–139, 1962.

[9] Tom Duff. Interval arithmetic and recursive subdivision for implicit functions and constructive solid geometry. In Edwin E. Catmull, editor, *Computer Graphics (SIGGRAPH '92 Proceedings)*, volume 26, pages 131–138, July 1992.

[10] E. Durand. Solutions numeriques des equations algebriques. *Tome I, Masson, Paris*, 1960.

[11] Herbert Edelsbrunner and Ernst Peter Mucke. Simulation of simplicity: A technique to cope with degenerate cases in geometric algorithms. *ACM Transactions on Graphics*, 9(1):66–104, 1990.

[12] Ioannis Z. Emiris and John F. Canny. A general approach to removing degeneracies. In *Proceedings of the 32nd IEEE Symposium on the Foundations of Computer Science*, pages 405–413, 1994.

[13] Michal Etzion and Ari Rappoport. Computing the Voronoi diagram of a 3-D polyhedron by separate computation of its symbolic and geometric parts. In *Proc. Symposium on Solid Modeling and Applications*, pages 167–178, 1999.

[14] Steven Fortune. Polyhedral modeling with exact arithmetic. *Proc. Symposium on Solid Modeling and Applications*, pages 225–234, 1995.

[15] Steven Fortune. Robustness issues in geometric algorithms. In M. C. Lin and D. Manocha, editors, *Applied Computational Geometry: Towards*

Geometric Engineering, volume 1148 of *Lecture Notes Comput. Sci.*, pages 9–14. Springer-Verlag, 1996.

[16] Steven Fortune and Christopher J. van Wyk. Static analysis yields efficient exact integer arithmetic for computational geometry. *ACM Transactions on Graphics*, 15(3):223–248, July 1996.

[17] LiDIA Group. A library for computational number theory. Technical report, TH Darmstadt, Fachbereich Informatik, Institut fur Theoretische Informatik, 1997.

[18] C. M. Hoffmann. The problems of accuracy and robustness in geometric computation. *IEEE Computer*, 22(3):31–41, March 1989.

[19] C.M. Hoffmann. *Geometric and Solid Modeling*. Morgan Kaufmann, San Mateo, California, 1989.

[20] Kenneth E. Hoff III, Tim Culver, John Keyser, Ming Lin, and Dinesh Manocha. Fast computation of generalized Voronoi diagrams using graphics hardware. In *Computer Graphics Annual Conference Series (SIGGRAPH '99)*, pages 277–286, 1999.

[21] M. Karasick, D. Lieber, and L. R. Nackman. Efficient Delaunay triangulations using rational arithmetic. *ACM Trans. Graph.*, 10(1):71–91, January 1991.

[22] John Keyser, Tim Culver, Dinesh Manocha, and Shankar Krishnan. MAPC: A library for efficient and exact manipulation of algebraic points and curves. In *Proc. 15th Annual ACM Symposium on Computational Geometry*, pages 360–369, 1999.

[23] John Keyser, Shankar Krishnan, and Dinesh Manocha. Efficient and accurate B-rep generation of low degree sculptured solids using exact arithmetic I: Representations and II: Computation. *Computer Aided Geometric Design*, 16(9):841–882, October 1999.

[24] Shankar Krishnan, Mark Foskey, John Keyser, Tim Culver, and Dinesh Manocha. PRECISE: Efficient multiprecision evaluation of algebraic roots and predicates for reliable geometric computation. Technical Report TR00-008, University of North Carolina-Chapel Hill, 2000.

[25] Shankar Krishnan and Dinesh Manocha. An efficient surface intersection algorithm based on the lower dimensional formulation. *ACM Transactions on Graphics*, 16(1):74–106, 1997.

[26] Shankar Krishnan, Dinesh Manocha, M. Gopi, Tim Culver, and John Keyser. BOOLE: A boundary evaluation system for boolean combinations of sculptured solids. *International Journal of Computational Geometry and Applications*, 11(1):105–144, 2001.

[27] Chen Li and Chee Yap. A new constructive root bound for algebraic expressions. In *Proceedings of the Symposium on Discrete Algorithms*, 2001.

[28] Chen Li and Chee Yap. Recent progress in exact geometric computation. http://www.cs.nyu.edu/exact/doc/dimacs.ps.gz, 2001.

[29] Victor Milenkovic. Robust construction of the Voronoi diagram of a polyhedron. In *Proc. 5th Canad. Conf. Comput. Geom.*, pages 473–478, 1993.

[30] A. A. G. Requicha and H. B. Voelcker. Boolean operations in solid modeling: boundary evaluation and merging algorithms. *Proceedings of the IEEE*, 73(1), 1985.

[31] Mark Segal. Using tolerances to guarantee valid polyhedral modeling results. In *Proceedings of ACM Siggraph*, pages 105–114, 1990.

[32] Jules Vleugels and Mark Overmars. Approximating generalized Voronoi diagrams in any dimension. Technical Report UU-CS-1995-14, Department of Computer Science, Utrecht University, 1995.

[33] Chee Yap. Towards exact geometric computation. In *Proc. 5th Canad. Conf. Comput. Geom.*, pages 405–419, 1993.

[34] Chee Yap and Thomas Dube. The exact computation paradigm. In D. Z. Du and F. Hwang, editors, *Computing in Euclidean Geometry*, pages 452–492. World Scientific Press, Singapore, 1995.

Chapter 9

FEATURE LOCALIZATION ERROR IN 3D COMPUTER VISION

Daniel D. Morris
Northrop Grumman Corp.
1501 Ardmore Blvd.
Pittsburgh, PA 15217, USA
Daniel_D_Morris@mail.northgrum.com

Takeo Kanade
Robotics Institute
Carnegie Mellon University
Pittsburgh, PA 15213, USA
tk@ri.cmu.edu

Abstract Uncertainty modeling in 3D Computer Vision typically relies on propagating the uncertainty of measured feature positions through the modeling equations to obtain the uncertainty of the 3D shape being estimated. It is widely believed that this adequately captures the uncertainties of estimated geometric properties when there are no large errors due to mismatching. However, we identify another source of error which we call feature *localization error*. This captures how well a feature corresponds to the true 3D point, rather than how well features correspond over multiple images. We model this error as independent of the tracking error, and when combined as part of the total error, we show that it is significant and may even dominate the 3D reconstruction error.

Keywords: Covariance estimation, Computer Vision, bias, feature localization error, uncertainty

Introduction

Feature tracking is a key step in 3D estimation for many Computer Vision algorithms. It is used in Stereo Vision, where it is typically called image registration, in Structure from Motion, and in other algorithms that estimate geometric properties of the world. The use of image points, which correspond to surface features of a 3D object, enables these methods to be formulated in purely geometric terms.

One part of estimating a geometric property of the world, such as a 3D position, consists in obtaining the accuracy or uncertainty of the estimate. Since the camera projection equations are generally known, the errors in the final 3D estimate are a direct result of the errors in the measured image positions of the features. Methods that model this 3D uncertainty typically start by assuming an image covariance around each feature point to model the image noise in finding the feature. Then this covariance uncertainty is propagated through the linearized projection equations to obtain a covariance of the final 3D estimate. This is the standard approach for uncertainty modeling in Computer Vision, and has been investigated for stereo by Matties and Shafer, 1987 and Kanatani, 1993, and in more general Structure from Motion by Weng et al., 1989; Young and Chellappa, 1992; Szeliski and Kang, 1994; Thomas et al., 1994; Daniilidis and Spetsakis, 1996; Haralick, 1996; Morris and Kanade, 1998; Morris et al., 2000 and others.

This uncertainty modeling is correct in the idealized case where feature position estimates are unbiased and the noise is small so first order terms dominate. As long as there are no mismatches, which create large error terms, it is widely assumed this model provides a good uncertainty of the final 3D estimate. Indeed we agree that this method is appropriate when symmetric markers are placed on an object such as in human figure tracking or in photogrammetry. However, we contend that when natural features, such as corners, are used, this uncertainty modeling scheme leads to overly optimistic uncertainty estimates. This is because it overlooks an important component of the error which we call the *feature localization error*.

In order to specify what we mean by feature localization error we will first distinguish it from *feature tracking error*, which is what most methods actually consider. Feature tracking error is the error involved in registering the same point, whatever the point is, in multiple images. This can be done manually or using template matching with sub-pixel accuracy, provided there is adequate texture. Feature localization error, on the other hand, corresponds to the error identifying the exact projection of the desired 3D point in the image. This occurs because the inherent resolution limitations of digital images prevent 3D feature points from being precisely identified in the first place.

Feature Localization Error in 3D Computer Vision

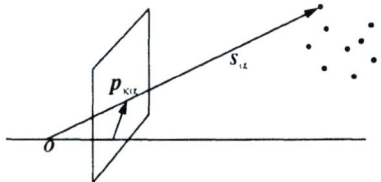

Figure 9.1. A 3D point s_α is projected onto the image point $p_{\kappa\alpha}$.

In this paper we identify the cases in which localization error ought to be modeled. We propose a method to model this localization error and show how it can be incorporated into standard uncertainty modeling schemes. Along the way we show that, contrary to common intuition, this effect cannot be captured by simply increasing the covariance of the tracking. Futhermore we show that the effects of localization error can be equal to or greater than the effects of errors in typical feature registration. Hence it should not be overlooked in 3D estimation schemes that include error models.

1. Standard Error Modeling

We begin by describing the standard error propagation method used in 3D Computer Vision to obtain uncertainty models, and then show how it may fail to model an important component of the error.

1.1 Parametric Modeling

We will assume a perspective camera projection model as illustrated in Fig. 9.1. A 3D point s_α is projected onto point $p_{\kappa\alpha}$ in the κth image. Our goal is to use a set of projections of multiple features in multiple images to estimate a parameter vector, θ, which may include 3D feature positions s_α and possibly camera motion parameters. We express our estimation task in vector format as:

$$p = \Pi(\theta), \qquad (9.1)$$

where vector p includes all the image feature positions, $p_{\kappa\alpha}$, and $\Pi(\cdot)$ encodes our known camera projection equations. We want to find the vector $\hat{\theta}$ that best explains our measured data, p.

1.2 Estimation and Uncertainty Modeling

Next we assume that the noise is small such that the first order terms dominate, and that the noise is unbiased. We can write:

$$p = \bar{p} + \Delta p, \qquad (9.2)$$

where $\bar{p}$ is the true measurement with no noise. The noise term Δp has the statistics:
$$E[\Delta p] = 0, \tag{9.3}$$
and
$$V_p = E[\Delta p \Delta p^\top], \tag{9.4}$$
where $E[\cdot]$ denotes expectation, $^\top$ is the transpose, and V_p is the data covariance.

To estimate our unknown parameter vector θ, we first define a cost function based on the difference between our projected model and our data:
$$J(\theta) = \frac{1}{2}(p - \Pi(\theta))^\top V_p^{-1}(p - \Pi(\theta)). \tag{9.5}$$

The weighted least squares estimate for $\hat{\theta}$ is solved using an appropriate algorithm such as Structure from Motion or Stereo, depending on the parameters included in θ. When $J(\theta)$ is minimized the solution is unbiased, to first order, and achieves the Cramer-Rao lower bound. Hence the covariance of the estimate, $\hat{\theta}$, is given by the inverse of the Fisher Information matrix[1] which is equal to the Hessian at the optimal solution:
$$V_\theta = (\nabla_\theta^2 J(\theta))^{-1}. \tag{9.6}$$

The covariance, V_θ, is our uncertainty model for the estimated parameter vector $\hat{\theta}$.

This is the standard approach for first order error modeling in Computer Vision. It considers only the feature matching or tracking error, and propagates this onto the 3D uncertainty model. One property to note is that when there is no error in registering features between images, there will be no error in 3D estimation.

2. Localization Error

The key point of this paper is to show that while the derivations in the previous section are correct, when that method is used to obtain uncertainty estimates for standard Computer Vision algorithms, it often significantly underestimates these uncertainties. We will illustrate this with a simple example.

2.1 Estimating a Cross Ratio

Figure 9.2 shows the first and last images of a sequence where four points have been identified in the first image and tracked using an affine tracker as done by Shi and Tomasi, 1994. The four points are collinear, except for noise, and in this example the component of the 3D shape we wish to estimate is the

Feature Localization Error in 3D Computer Vision

Figure 9.2. The first and last images in an image sequence of windows on a wall. Four collinear corners are marked, and our goal is to estimate the cross ratio, R, of these points.

Figure 9.3. Uniform noise is assumed for all features, and fitted points are constrained to be collinear as well as all images must have the same cross ratio, R.

cross ratio of these points given by:

$$R = \frac{(a-b)(c-d)}{(a-d)(b-c)}, \qquad (9.7)$$

where a, b, c, and d are the positions of the points on the line. The tracked features, p, are assumed to be equal to their true values plus uniform, uncorrelated Gaussian noise. Hence the covariance, V_p, is given by $\epsilon^2 I$. The magnitude of the pixel noise, ϵ^2, may be known from the feature tracker, or else can be calculated as in Kanatani and Morris, 2001 and Morris, 2001.[2]

Our minimal parametrization, θ, includes as its first term the cross ratio, R, followed by two line parameters for each image, and three position parameters for the points on thes lines in each image. The fourth point in each image can be specified by these three points and the cross ratio. The projection function, $\Pi(\theta)$, transforms these parameters to a set of coordinates for the four collinear points in each image. All these parameters including R are optimally estimated by minimizing $J(\theta)$ defined in Eq. (9.5). The variance of R due to tracking error, σ_T^2, is obtained by selecting the top left element of V_θ in Eq. 9.6.

Each image can be thought of as giving an estimate for R, and in the limit of infinite images the uncertainty due to tracking, σ_T^2, will approach zero, and presumeably R will be perfectly estimated. The results for estimating R using an increasing number of images is shown in Fig. 9.4 (a). We see that indeed, as more images are used, the standard deviation of our estimate decreases. But

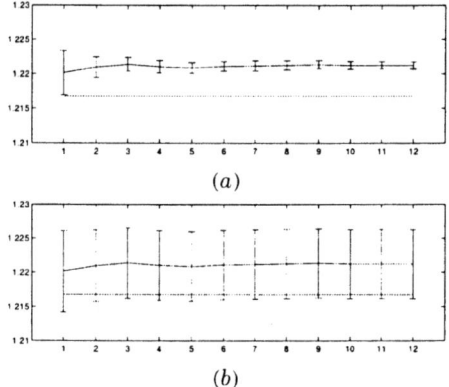

Figure 9.4. (a) Estimates of the cross ratio R using the first n images of the sequence in Fig. 9.2, where n is shown on the abscissa. The true value is shown by the horizontal line. The error bars in (a) indicate the standard deviation for the tracking error, and in (b) give the standard deviation when localization error is included.

unfortunately, as more images are used the accuracy of our estimate does not increase in corresponding measure to the standard deviation. We obtained the true value of R by measuring the distances between the corners on the actual windows before taking photos.

2.2 Modeling Localization Error

Evidently there is an additional source of error beyond purely registering features in the images. Consider the case in which the features had been perfectly registered: our estimate would have no variance, and yet it is unlikely that it would correspond to the true cross ratio of the windows. The additional source of error must come from identifying the exact 2D projections of the windows' corners in the first image. Thus we propose that there are two main sources of error: a tracking error involved in registering the same point in multiple images, and a localization error which captures the error in identifying the desired 3D point in a 2D image. We will model these as independent sources of error, and thus can write the total variance as:

$$\sigma_R^2 = \sigma_T^2 + \sigma_L^2. \tag{9.8}$$

The first term on the right is the tracking variance, and the second term is the localization variance. This is illustrated in Figure 9.5.

Next we would like derive an approximation for the localization error. Let i identify the image in which the features are first identified. We assume that we initially located the points with accuracy $\Delta \boldsymbol{p}_i$ in this image. Then we ask what is the corresponding perturbation of the fitted parameters: $\Delta \boldsymbol{\theta}_i$ containing R and the other parameters in this image? It is straightforward to let p_i be our data in Equation 9.1 and expand it to obtain:

$$\Delta p_i = (\nabla_\theta \Pi(\theta))^T \Delta \theta_i, \qquad (9.9)$$

where $\nabla_\theta \Pi(\theta)$ is the appropriate gradient matrix of $\Pi(\theta)$. Since this equation over-constrains the θ_i, it can be inverted and the expectation of the square taken to give us:

$$V_{\theta i} = E[\theta_i \theta_i^T] = \left((\nabla_\theta \Pi(\theta))^T\right)^\dagger V_{p_i} (\nabla_\theta \Pi(\theta))^\dagger, \qquad (9.10)$$

where "†" denotes the pseudo-inverse. The localization variance, σ_L^2, is then the top left element of $V_{\theta i}$. Figure 9.4 (b) shows the effect on the uncertainty when σ_L^2 is included. It models what might otherwise be called a bias in our estimate.

2.3 Localization Error in 3D

In 3D Computer Vision we often want to estimate the positions of particular points on an object, or distances between known points on an object, such as the object width. Past work has obtained variances for these estimates but only using tracking error. There will, however, also be a localization component to the error.

We model the localization error being generated in the first image as illustrated in Fig. 9.6. A true 3D point, s, projects into the image at position p. Due to the difficulty in exactly identifying this point in the image, as illustrated in Fig. 9.5, an image localization error, Δp_L, will be added with standard deviation σ_L. Now the point identified, p_L, corresponds to a 3D point s_L, which we assume will be the point that is reconstructed. We would like to know what the distribution of Δs_L is with respect to the true point s. This will depend on the surface properties of the object that is being tracked, but for simplicity, we assume that Δs_L will have uniform standard deviation of magnitude σ_{sL} around point s. Using a perspective image projection of this onto Δp_L we deduce:

$$\sigma_L = \sigma_{sL} \frac{f}{Z}, \qquad (9.11)$$

where f is the focal length and Z is the depth of the point in 3D. If we know the image localization error, σ_L, we can find the feature localization error using this equation. We assume that we can only localize a feature to pixel accuracy, as illustrated in Fig. 9.5, and hence that σ_L is 1 pixel.

Now after 3D estimation is performed by minimizing $J(\theta)$, we will obtain the 3D positions s for each point, and their tracking covariances V_{sT}. Then to obtain the full covariance for each point we simply add the localization

114

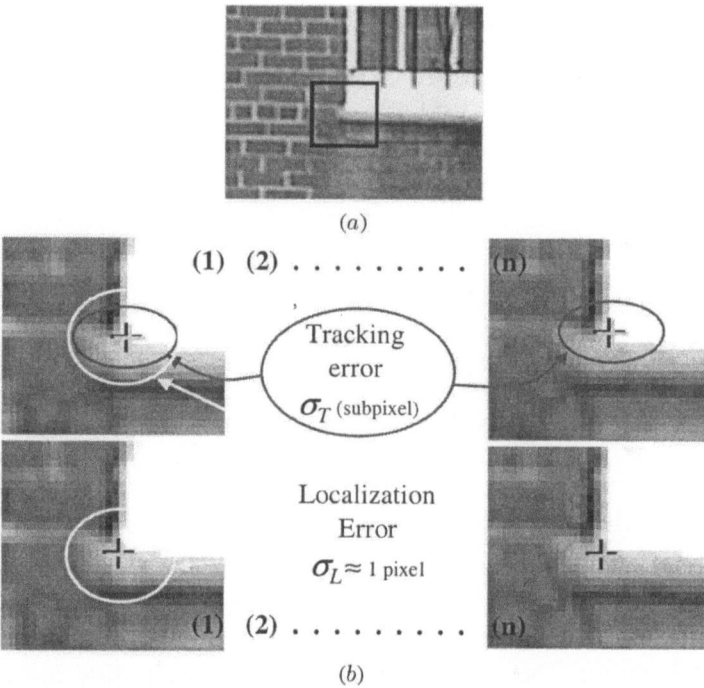

Figure 9.5. We would like to find the image position corresponding to the corner of the window shown in (a), and the corresponding positions in all the images. We see from the close-up views that it is not easy to identify the exact coordinates of the corner. The left two figures in (b) show possible positions of the corner in the first image. If either of these are chosen, then the corresponding point is obtained in the last image shown on the right. The tracking error specifies how accurately the point chosen in the first image corresponds to points in subsequent images. The localization error specifies how accurately the point chosen in the first image corresponds to the projection of the actual 3D point.

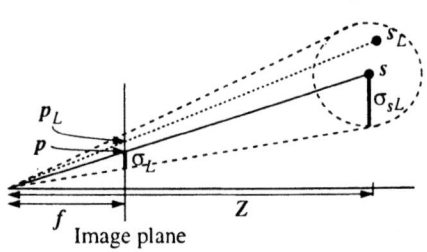

Figure 9.6. Due to localization error, point s_L is tracked and reconstructed, rather than the true point s. We assume a one pixel image localization error, σ_L, and from this estimate the 3D localization error, σ_{sL}, as in Eq. 9.11.

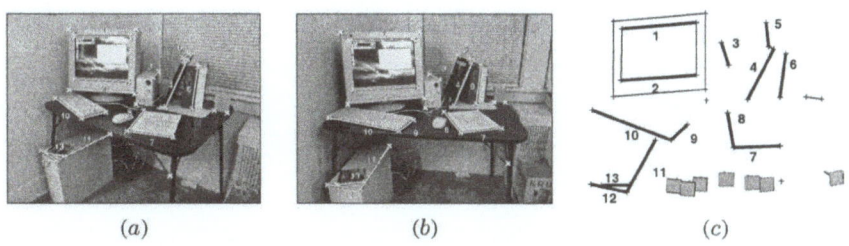

Figure 9.7. (a) The first image and (b) the last image from a seven image sequence of a computer and desk. (c) The 3D reconstruction.

covariance:
$$V_s = V_{sT} + \sigma_{sL}^2 I, \qquad (9.12)$$

where I is the 3×3 identity matrix. This is our completed error model.

3. Experiment

We illustrate the significance of localization error on a real example. Seven images of a desk were taken from unknown camera positions and the 3D geometry was reconstructed as shown in Fig. 9.7. Our goal was to estimate the width of the computer monitor, line 1, but the reconstruction is only up to a scale factor. We can fix the scale using any one of the marked lines. The resulting estimates for the length of line 1 are given in Fig. 9.8(a), along with their standard deviations calculated from the tracking uncertainty. If the tracking uncertainty correctly modeled the true error, then we would expect 95% of these estimates to be within two standard deviations of the true value. This is clearly not the case. However, if we add the localization error estimates as in Eq. 9.8, then we obtain Fig 9.8(b). The standard deviations are significantly larger now, and so provide a better model for the error.

4. Conclusion

Our goal is not a more accurate 3D estimate, but a better model of the uncertainty of the estimated 3D points. If we simply wanted to estimate arbitrary points on the surface of an object or to estimate the camera motion parameters, then these estimates would not be affected by localization error. However, we have shown that when we wish to estimate the position of *specified* 3D points in the world, then localization error affects the estimation accuracy. We developed a model for localization error, and showed empirically that its effect is significant. Unlike tracking error, the effect of localization error is not reduced as more images are taken and so when many images are used it may dominate, and hence the widely held assumption that using registration error alone in

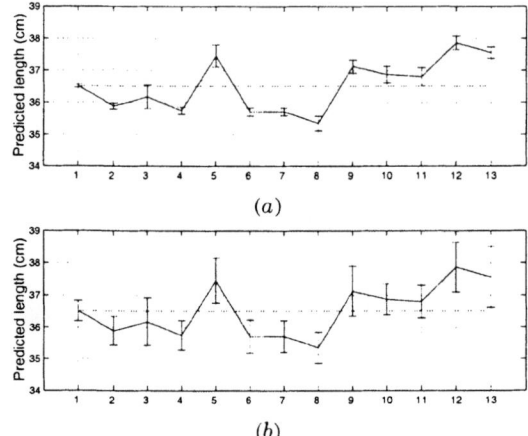

Figure 9.8. The predicted monitor width is shown, along with its uncertainty in standard deviations for each of the thirteen lines. (*a*) Uncertainy is calculated using Eq. 9.6. (*b*) Localization error is included as in Eq. 9.8 assuming $\sigma_L = 1$ pixel for all features.

obtaining uncertainy of 3D shape must be revised. The cross ratio example shows that one cannot simply increase the registration covariances and hope that this will adequately model the localization error.

We modeled localization error as independent of tracking error. In order for this assumption to be valid, the appearance of a feature must not change too much. This is the case when camera motion is relatively small over the time when a feature is tracked, and this is true for many real examples of stereo and Structure from Motion.

An alternative interpretation of our results is that feature localization error provides a statistical model for a source of bias in Computer Vision problems that has been ignored previously.

Notes

1. We are assuming that the Fisher Information matrix is full rank. If it is not, then we must choose gauge constraints as described in Kanatani and Morris, 2001; Morris, 2001.

2. The covariance magnitude, ϵ^2, is calculated by noting that $2J(\theta)$ is a χ^2 variable whose expectation is equal to its number of degrees of freedom. If initially we let $V_p = I$, and we take our resulting cost, after fitting, to be equal to $E[J(\theta)]/\epsilon^2$, then we obtain ϵ^2.

References

Daniilidis, K. and Spetsakis, M. E. (1996). *Visual Navigation*. Lawrence Erlbaum Associates, Hillsdale, NJ. Chapter in book edited by Y. Aloimonos.

Haralick, R. M. (1996). Propagating covariance in computer vision. *Int. J. of Patt. Recog. and AI.*, 10(5):561–572.

Kanatani, K. (1993). Unbiased estimation and statistical analysis of 3-D rigid motion from two views. *IEEE Trans. Patt. Anal. Mach. Intell.*, 15(1):37–50.

Kanatani, K. and Morris, D. D. (2001). Gauges and gauge transformations for uncertainty description of geometric structure with indeterminacy. *IEEE Transactions on Information Theory*, 47(5):2017–2028.

Matties, L. and Shafer, S. A. (1987). Error modeling in stereo navigation. *IEEE Journal of Robotics and Automation*, RA-3(3):239–48.

Morris, D. D. (2001). *Gauge Freedoms and Uncertainty Modeling for 3D Computer Vision*. PhD thesis, Robotics Institute, Carnegie Mellon University, Pittsburgh, PA. (see also: CMU-RI-TR-01-15).

Morris, D. D. and Kanade, T. (1998). A unified factorization algorithm for points, line segments and planes with uncertainty models. In *Proc. Sixth Int. Conf. Comp. Vision*, pages 696–702, Bombay, India.

Morris, D. D., Kanatani, K., and Kanade, T. (2000). Uncertainty modeling for optimal structure from motion. In *Vision Algorithms: Theory and Practice*, pages 200–217, Berlin. Springer.

Shi, J. and Tomasi, C. (1994). Good features to track. In *Proc. Comp. Vision Patt. Recog.*, pages 593–600, Seattle.

Szeliski, R. and Kang S. B. (1994). Recovering 3D shape and motion from image streams using non-linear least squares. In *J. of Visual Comm. and Image Rep.*, **5**(1), pages 10–28.

Thomas, J. I., Hanson, A., and Oliensis, J. (1994). Refining 3D reconstructions: A theoretical and experimental study of the effect of cross-correlations. *CVGIP*, 60(3):359–370.

Weng, J., Huang, T., and Ahuja, N. (1989). Motion and structure from two perspective views: Algorithms, error analysis, and error estimation. *IEEE Trans. Patt. Anal. Mach. Intell.*, 11(2):451–476.

Young, G.-S. J. and Chellappa, R. (1992). Statistical analysis of inherent ambiguities in recovering 3-D motion from a noisy flow field. *IEEE Trans. Patt. Anal. Mach. Intell.*, 14(10):995–1013.

Chapter 10

BAYESIAN ANALYSIS OF COMPUTER MODEL OUTPUTS

Jeremy Oakley, Anthony O'Hagan
Department of Probability and Statistics
University of Sheffield
Houndsfield Road
Sheffield S3 7RH
United Kingdom
{j.oakley,a.ohagan}@sheffield.ac.uk

Abstract   We consider various statistical problems associated with the use of complex deterministic computer models. In particular, we focus on exploring the uncertainty in the output of the model that is induced by uncertainty in some or all of the model input parameters. In addition, we consider the case when the computer model is computationally expensive, so that it is necessary to be able to describe the output uncertainty based on a small number of runs of the model itself.

Keywords: Deterministic computer model, uncertainty analysis, sensitivity analysis, Gaussian process

1. Computer models and statistics

Computer models are used in a variety of scientific fields. In some applications they may be used to predict future events, for example in weather or economic forecasting. When experimentation on real life systems is too costly or impractical, a computer model of the system may be used as a surrogate. An example of this scenario would be investigating the dispersal of a pollutant from a source under a variety of atmospheric conditions. We think of the computer model as returning a number of outputs when provided with a set of inputs that will correspond to the situation being modelled. Typically, these models are deterministic, so that running them repeatedly at the same inputs will always give the same outputs. We describe the process of running a computer model at a variety of different input values as a 'computer experiment'.

Despite the purely deterministic nature of these models, there are various problems regarding the use of these models that are of interest to statisticians. This is because the user of the model will invariably encounter uncertainty at some stage of the computer experiment. Here we identify four areas where statisticians can make a contribution:

Uncertainty analysis.

When the user applies their model to any specific situation, they must choose actual values to use for their model input parameters. But what if the input parameter represents a quantity that is impossible or impractical to measure? In most cases there will be uncertainty regarding the values of some or even all of the input parameters that should be used in any single application. The user could simply plug in their best guesses for the inputs, but the input uncertainty could have the effect of inducing considerable uncertainty in the output; if the 'true' input values are unknown, the output of the model evaluated at the 'true' inputs must also be unknown. The consequences of this input uncertainty need investigating.

Sensitivity analysis.

This is concerned with studying the influence of individual input parameters (or combinations of inputs) on the model. Where there is uncertainty about the input values, it may be of particular interest to identify which uncertain inputs are the most influential, in some sense, in inducing the overall output uncertainty.

Model inadequacy and calibration.

So far we have only considered the model itself, and not the real-world process the model is attempting to describe. Two related issues arise when observations from the real-world process are available. Firstly, the model will invariably not be describing the true process perfectly, though it will of course still give predictions at parts of the input space still unobserved in reality. How should these two sources of information, the model and the real observations be combined? Secondly, can the real observations be used to tune the input parameters of the model so that it is more accurate, i.e can we calibrate the model to reality?

Experimental design and model interpolation.

All the issues described so far can be complicated by computer models that are computationally expensive, so that running the model at any single set of input values takes a non-trivial amount of computing time. If it is only practical to run the model at a small number of distinct sets of inputs, we may be forced to make predictions of the model output itself at other untested input values. Secondly, the choice of input values at which we run the model will become important.

In this paper we will concentrate on uncertainty and sensitivity analysis for computationally expensive models. Suggested references for model inadequacy and calibration are Kennedy and O'Hagan, 2001 and Craig et al., 2001, and Sacks et al., 1989 for experimental design. In the next section we first describe the framework for handling computationally expensive models. In sections three and four we deal with uncertainty and sensitivity analysis, and illustrate each with an example.

2. Bayesian inference about functions using Gaussian processes

We first denote the computer model by $y = \eta(\mathbf{x})$, where $\mathbf{x}$ are the model inputs and y is the model output. Though the output will typically be a vector, we only consider a scalar output here.

In the Bayesian approach, we begin by treating the function $\eta(.)$ as a random variable. We consider $\eta(.)$ to be random simply in the sense that it is unknown, as opposed to being the product of some random process. The computer model can tell us the value of $\eta(\mathbf{x})$ exactly, but until the code is run, this quantity is unknown, and so we can describe our beliefs about $\eta(\mathbf{x})$ through a probability distribution. Thus we first need to describe our prior beliefs about $\eta(.)$. We now describe the Gaussian process model for an unknown function $\eta(.)$. Gaussian processes have been used before for modelling computer codes, and examples are Currin et al., 1991 and Haylock and O'Hagan, 1996. The key requirement is that $\eta(.)$ is a smooth function, so that if we know the value of $\eta(\mathbf{x})$, we should have some idea about the value of $\eta(\mathbf{x}')$ for $\mathbf{x}$ close to $\mathbf{x}'$. It is this property of $\eta(.)$ that will allow us to deal with computationally expensive models; by considering the full information about $\eta(.)$ that we obtain from each single run, we can reduce the number of runs needed.

We represent our uncertainty about $\{\eta(\mathbf{x}_1), \ldots, \eta(\mathbf{x}_n)\}$ for any set of points $\{\mathbf{x}_1, \ldots, \mathbf{x}_n\}$ through a multivariate normal distribution. The mean of $\eta(\mathbf{x})$ is given by

$$E\{\eta(\mathbf{x})|\beta\} = \mathbf{h}(\mathbf{x})^T \beta, \qquad (10.1)$$

conditional on β. The vector $\mathbf{h}(.)$ consists of q known regression functions of $\mathbf{x}$, and β is a vector of coefficients. The choice of $\mathbf{h}(.)$ is arbitrary, though it should be chosen to incorporate any beliefs we might have about the form of $\eta(.)$. The covariance between $\eta(\mathbf{x})$ and $\eta(\mathbf{x}')$ is given by

$$Cov\{\eta(\mathbf{x}), \eta(\mathbf{x}')|\sigma^2\} = \sigma^2 c(\mathbf{x}, \mathbf{x}'), \qquad (10.2)$$

conditional on σ^2, where $c(\mathbf{x}, \mathbf{x}')$ is a function which decreases as $|\mathbf{x} - \mathbf{x}'|$ increases, and also satisfies $c(\mathbf{x}, \mathbf{x}) = 1 \; \forall \mathbf{x}$. The function $c(., .)$ must ensure that the covariance matrix of any set of outputs $\{\eta(\mathbf{x}_1), \ldots, \eta(\mathbf{x}_n)\}$ is positive semi definite. A typical choice is

$$c(\mathbf{x}, \mathbf{x}') = \exp\{-(\mathbf{x} - \mathbf{x}')^T B (\mathbf{x} - \mathbf{x}')\}, \quad (10.3)$$

where B is a diagonal matrix of (positive) roughness parameters. Conventionally, a weak prior of β and σ^2 in the form $p(\beta, \sigma^2) \propto \sigma^{-2}$ is used. In Oakley, 2002 a means of including proper prior information about the function $\eta(.)$ is presented, through the use of the conjugate prior, the normal inverse gamma distribution. For chosen values of the hyperparameters $a, d, \mathbf{z}$ and V we have

$$p(\beta, \sigma^2) \propto (\sigma^2)^{-\frac{1}{2}(d+q+2)} \exp[-\{(\beta - \mathbf{z})^T V^{-1} (\beta - \mathbf{z}) + a\}/(2\sigma^2)]. \quad (10.4)$$

(Recall that q is the number of regressors in the mean function).

The output of $\eta(.)$ is observed at n design points, $\mathbf{x}_1, \ldots, \mathbf{x}_n$ to obtain data $\mathbf{y}$. Given the prior in (10.4) it can be shown that

$$\frac{\eta(\mathbf{x}) - m^*(\mathbf{x})}{\hat{\sigma}\sqrt{c^*(\mathbf{x}, \mathbf{x})}} | \mathbf{y}, B \sim t_{d+n}, \quad (10.5)$$

where

$$m^*(\mathbf{x}) = \mathbf{h}(\mathbf{x})^T \hat{\beta} + \mathbf{t}(\mathbf{x})^T A^{-1} (\mathbf{y} - H\hat{\beta}), \quad (10.6)$$

$$c^*(\mathbf{x}, \mathbf{x}') = c(\mathbf{x}, \mathbf{x}') - \mathbf{t}(\mathbf{x})^T A^{-1} \mathbf{t}(\mathbf{x}') +$$
$$(\mathbf{h}(\mathbf{x})^T - \mathbf{t}(\mathbf{x})^T A^{-1} H)(H^T A^{-1} H)^{-1}$$
$$\times (\mathbf{h}(\mathbf{x}')^T - \mathbf{t}(\mathbf{x}')^T A^{-1} H)^T. \quad (10.7)$$

$$\mathbf{t}(\mathbf{x})^T = (c(\mathbf{x}, \mathbf{x}_1), \ldots, c(\mathbf{x}, \mathbf{x}_n)), \quad (10.8)$$

$$H^T = (\mathbf{h}^T(\mathbf{x}_1), \ldots, \mathbf{h}^T(\mathbf{x}_n)), \quad (10.9)$$

$$A = \begin{pmatrix} 1 & c(\mathbf{x}_1, \mathbf{x}_2) & \cdots & c(\mathbf{x}_1, \mathbf{x}_n) \\ c(\mathbf{x}_2, \mathbf{x}_1) & 1 & & \vdots \\ \vdots & & \ddots & \\ c(\mathbf{x}_n, \mathbf{x}_1) & \cdots & & 1 \end{pmatrix}, \quad (10.10)$$

$$\hat{\beta} = V^*(V^{-1}\mathbf{z} + H^T A^{-1}\mathbf{y}), \quad (10.11)$$

$$\hat{\sigma}^2 = \{a + \mathbf{z}^T V^{-1} \mathbf{z} + \mathbf{y}^T A^{-1} \mathbf{y} - \hat{\beta}^T (V^*)^{-1} \hat{\beta}\}$$
$$\times (n + d - 2)^{-1}, \quad (10.12)$$

$$V^* = (V^{-1} + H^T A^{-1} H)^{-1} \quad (10.13)$$

$$\mathbf{y}^T = (\eta(\mathbf{x}_1), \ldots, \eta(\mathbf{x}_n)). \quad (10.14)$$

Full details of the prior to posterior analysis can be found in O'Hagan, 1994. In this paper we will simply condition on a posterior estimate of B, rather than taking into account the uncertainty that we may have. The estimate could be the posterior mode, though with a small sample of data the likelihood can be fairly flat. Cross-validation procedures can be effective; leave each observation out in turn and choose B to minimise the error between the posterior mean of the omitted output and the known true value. Ideally we should be allowing for the uncertainty in B, though Kennedy and O'Hagan, 2001 have suggested that this uncertainty may not be important. Neal, 1999 uses MCMC sampling to sample from the posterior distribution of B.

3. Uncertainty Analysis

We illustrate the concept of uncertainty analysis with the following simple example. Suppose there has been an accidental release of radionuclides from a point source, and we wish to predict the resulting concentration of radionuclides at a particular location near to the source. Mathematical models have been developed to perform such a task, for example, the Gaussian Plume Diffusion model in Clarke, 1979. The model assumes that the dispersion of radionuclides in the horizontal and vertical directions can be described by Gaussian distributions, and the simplest version of the model is given by

$$C(x,y,z) = \frac{Q}{2\pi u_{10} \sigma_z \sigma_y} \exp\left[-\frac{1}{2}\left\{\frac{y^2}{\sigma_y^2} + \frac{(z-h)^2}{\sigma_z^2}\right\}\right], \quad (10.15)$$

where C is the air concentration of the radionuclide, q is the total amount released, u_{10} is the wind speed at $10m$ above ground, σ_y and σ_z are the standard deviations of the horizontal and vertical Gaussian distributions respectively, h is the release height, and (x, y, z) are the coordinates along the wind direction, cross wind and above ground respectively. If we know the values of the inputs relating to a particular release, we can then make a prediction of the concentration at any location using this model. However, in practice it is quite likely that we will not know the values of all the inputs. It may not be possible to obtain a measurement of the intial quantity released, Q, or the wind speed u_{10} may be unknown. If we do not know the true values of all the inputs, we will not know the value of the output of the model evaluated at the true inputs. Consequently, this output, which we consider to be the 'true' output, is a random variable. The aim of uncertainty analysis is to learn about the uncertainty induced in the true output by the uncertainty in the inputs.

Since this model is only an approximation of the real life process, it is unlikely that if we do know the true values of all the input parameters, the output of the model will be the same as the observed concentration in reality. In addition, it may not always be meaningful to talk about a 'true' value of an input parameter. For example, we can think of the true value of Q as being the precise quantity of a radionuclide released during a particular accident. However, if we consider the two standard deviation parameters σ_y and σ_z, then if the real behaviour of the plume of radionuclides is not perfectly described by

Gaussian distributions, then what do the 'true' values of these parameters represent? These issues are not considered in uncertainty analysis. In the case of predicting some event in the future, the computer model may be the only source of information available, and the uncertainty resulting from unknown inputs needs to be understood. A model may give a very accurate prediction given all the correct inputs, but still be rendered ineffective if uncertainty about a particular input value results in high uncertainty about the output. A decision to invest resources in learning more about model inputs can be guided by an uncertainty analysis.

For a general model $y = \eta(\mathbf{x})$, we denote the true unknown inputs by $\mathbf{X}$, so that the corresponding true unknown output is $Y = \eta(\mathbf{X})$. The first step is for the model user to consider their uncertainty about $\mathbf{X}$, and represent this by a probability distribution $G(\mathbf{x})$. This in itself is a major task, though we do not consider the elicitation of $G(\mathbf{x})$ here. Once $G(\mathbf{x})$ is specified, this induces a distribution on Y, and the distribution of Y is known as the uncertainty distribution. Uncertainty analysis can then be thought of as making inferences about the uncertainty distribution.

In principle the uncertainty distribution can be obtained by Monte Carlo methods; draw a large sample of inputs $\mathbf{x}_1, \ldots, \mathbf{x}_n$ from $G(\mathbf{x})$, and evaluate $y_1 = \eta(\mathbf{x}_1), \ldots, y_n = \eta(\mathbf{x}_n)$ to obtain a sample from the distribution of Y. This is impractical for computationally expensive functions $\eta(.)$, as n will need to be large.

Haylock and O'Hagan, 1996 use the Gaussian process model to make inferences about the mean and variance of Y, and Oakley and O'Hagan, 2000 make inferences about the distribution and density functions of Y.

If we consider the distribution function

$$F(s) = \int_{\mathcal{X}} I\{\eta(\mathbf{x}) \leq s\} dG(\mathbf{x}), \qquad (10.16)$$

where $\mathcal{X}$ is the sample space of $\mathbf{X}$ and $I\{.\}$ denotes the indicator function, we can see that since $F(s)$ is a function of $\eta(.)$, and we are treating $\eta(.)$ as a random variable, then $F(s)$ must also be a random variable. In Oakley and O'Hagan, 2000 we describe in detail how to make inferences about any functional of $\eta(.)$ using a simulation procedure. In essence there are two steps to this procedure. The first step is to generate a random function, denoted by $\eta_{(i)}(.)$, from the posterior distribution of $\eta(.)$, which is given in (10.5). The function $\eta_{(i)}(.)$ is computationally cheap, and so the second step is to determine $F(s)|\eta(.) = \eta_{(i)}(.)$, which we denote by $F_{(i)}(s)$ using Monte Carlo sampling. Repeating these two steps gives us a sample $F_{(1)}(s), \ldots, F_{(N)}(s)$ from the distribution of $F(s)$.

3.1 Example: the ^{131}I model

We test these ideas with a computer code that models the behaviour of radioactive iodine in the human body. The model predicts the committed effective dose equivalent (CEDE), a measure of detriment over a fifty year period after the exposure, following the ingestion of a unit quantity of radioactive iodine. The iodine accumulates in the thyroid gland, and there are two unknown parameters in the model: the mass of the thyroid gland (w), and the fraction of iodine absorbed by the thyroid (f). Following a study by Dunning et al., 1981, lognormal input distributions are used. We have $\log w \sim N\left(2.889, 0.463^2\right)$ and $\log f \sim N\left(-1.315, 0.355^2\right)$. Independence is assumed between $\log w$ and $\log f$, and this completes the specification of $G(.)$. We write $\mathbf{x} = (\log w, \log f)$ and set $\mathbf{h}(\mathbf{x})^T = (1, \log w, \log f)$. For $c(\mathbf{x}, \mathbf{x}')$ we use

$$c(\mathbf{x}, \mathbf{x}') = \exp\left\{-(\mathbf{x} - \mathbf{x}')^T \begin{pmatrix} b_1 & 0 \\ 0 & b_2 \end{pmatrix} (\mathbf{x} - \mathbf{x}')\right\}.$$

Note that the transformation to the log scale has implications for the correlation structure; the correlation function is now only isotropic on the log scale. Haylock, 1997 believed this to be appropriate for the ^{131}I model, as correlations were expected to be smaller at inputs near the origin We evaluate the function $\eta(.)$ at nine points, and estimate b_1 and b_2 by their posterior mode, as in Haylock, 1997.

The pointwise median of distribution function is determined using the simulation procedure, and is plotted in figure 10.1 (solid line). We can also report posterior uncertainty about the distribution function. The dotted lines show pointwise 95% intervals for $F(s)$. These show how our estimate of $F(s)$ might change if we were to obtain further runs of the ^{131}I model. The ^{131}I model is not computationally expensive, and so we are able to calculate the true distribution function, based on 100,000 algorithm evaluations, which is shown by the dots. We have written 'dose' as the CEDE multiplied by 10^8. It can be seen that our estimate is very accurate using just the nine observations.

Suppose a dose in the region of 9 was considered critical. We can see from figure 10.1 that the probability that the output of the model, if run at the true inputs, would exceed 9 is small, so in this scenario, input uncertainty has not induced significant output uncertainty. Alternatively, suppose a dose in the region of 2 was considered critical. Now input uncertainty has made us much less certain whether or not the true output will exceed 2. In this case it may be necessary for the model user to attempt to reduce their input uncertainty.

4. Sensitivity Analysis

In sensitivity analysis, the aim is to learn about the influence of the individual input variables in the model. Local sensitivity analysis is concerned with

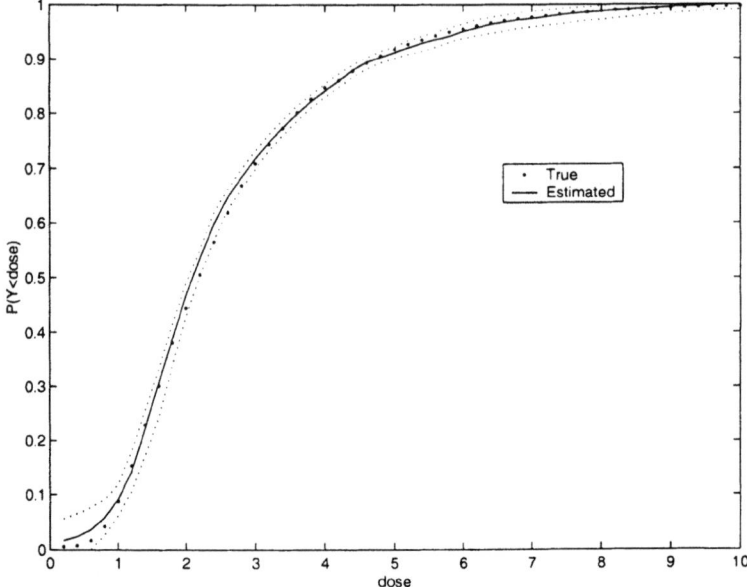

Figure 10.1. The estimated (solid line) and true (dots) distribution functions. The dotted lines represent pointwise 95% bounds for the distributions funtion.

derivatives of the output with respect to the various input variables at a particular point $\mathbf{x}_0$ in the input space. Where there is uncertainty about the true input $\mathbf{X}$, the model user will typically be interested in more than just infinitesimal variations of the input from a single point $\mathbf{x}_0$. Writing $\mathbf{X} = \{X_1, \ldots, X_d\}$, the user may wish to know the effect of their uncertainty about a single true input variable X_i on Y. Studying the effect of an input variable on the output over some interval for the input is known as global sensitivity analysis.

Saltelli et al., 2000 advocate global sensitivity analysis based on partitioning the variance of Y. The basic idea is to consider how individual inputs (and interactions between inputs) contribute to the uncertainty in Y. By looking at contributions of uncertain inputs to the variance of Y, there is an implicit assumption that the model user wishes to estimate Y, and has a quadratic loss function for the size of the estimation error. The user's expected loss is then given by $Var(Y)$. Now consider a particular input variable X_i, and suppose the user decides to learn the value of X_i before estimating Y. Ignoring the cost of observing X_i, the expected loss conditional on X_i will be $Var(Y|X_i)$, and so before X_i is actually observed, the expected loss is $E_{X_i}\{Var(Y|X_i)\}$. Since

$$Var(Y) = E_{X_i}\{Var(Y|X_i)\} + Var_{X_i}\{E(Y|X_i)\}, \qquad (10.18)$$

the expected reduction in loss is $Var_{X_i}\{E(Y|X_i)\}$. Hence the fraction

$$Var_{X_i}\{E(Y|X_i)\}/Var(Y) \qquad (10.19)$$

can be seen as a measure of the importance of the input variable X_i. We will refer to the term $S_i = Var_{X_i}\{E(Y|X_i)\}/Var(Y)$ as the main effect of input X_i. A large value of S_i indicates that there are potential benefits in reducing our uncertainty about X_i, since we would expect this to lead to a significant reduction in uncertainty about Y.

In the case of independent elements of $\mathbf{X}$, we can decompose the variance of Y into terms relating to the main effects and various interactions between the input variables. An ANOVA-like decomposition is given in Cox, 1982:

$$Var(Y) = \sum_{i=1}^{d} V_i + \sum_{i<j} V_{i,j} + \sum_{i<j<k} V_{i,j,k} + \ldots + V_{1,2,\ldots,d}, \qquad (10.20)$$

where

$$V_{i,j,k,\ldots} = Var_{X_i,X_j,X_k,\ldots}(Z_{i,j,k,\ldots}) \qquad (10.21)$$

$$Z_i = E(Y|X_i) \qquad (10.22)$$

$$Z_{j,k} = E\left(Y - \sum_{i=1}^{d} Z_i | X_j, X_k\right) \qquad (10.23)$$

$$Z_{k,l,m} = E\left(Y - \sum_{i=1}^{d} Z_i - \sum_{i<j} Z_{i,j} | X_k, X_l, X_m\right), \qquad (10.24)$$

and so on. Equation (10.20) gives us a partition of the expected loss, so for example the expected loss due solely to the interaction between inputs X_i and X_j is given by the term

$$V_{i,j} = Var_{X_i,X_j}\{E(Y|X_i,X_j) - E(Y|X_i) - E(Y|X_j)\}. \qquad (10.25)$$

This decomposition relies on the input variables being independent. For correlated inputs, we could first transform the inputs so that they are independent, though we may have problems when reporting the results back to the model user so that they are interpretable. Another possibility is to divide the inputs into groups so that there is independence between groups if not within groups. We can then partition the variance into terms relating to the main effects and interactions of the various groups.

Again, these terms are functions of $\eta(.)$, so for computationally expensive models where we treat $\eta(.)$ as a random variable, we estimate these terms based on a small number of model runs from the Gaussian process model for $\eta(.)$.

Details are given in Oakley and O'Hagan, 2002. Though these terms can be estimated by Monte Carlo sampling, typically large numbers of model runs are needed, and so Monte Carlo methods are impractical for computationally expensive models. In Oakley and O'Hagan, 2002, means for further exploring the influence of input variables are discussed, for example, to what extent the variance of Y is due to non-linearity in the inputs.

4.1 Example

We illustrate this methodology with a synthetic example. The following test function is used:

$$\eta(\mathbf{x}) = \mathbf{a}_1^T \mathbf{x} + \mathbf{a}_2^T \sin(\mathbf{x}) + \mathbf{a}_3^T \cos(\mathbf{x}) + \mathbf{x}^T M \mathbf{x} \qquad (10.26)$$

A fifteen dimensional input vector $\mathbf{x}$ is considered. The elements of the unknown true input $\mathbf{X}$ all have independent $N(0, 1)$ distributions. The weights $\mathbf{a}_1$, $\mathbf{a}_2$ and $\mathbf{a}_3$ are chosen so that one group of five input variables accounts for the majority of the variance of $Y = \eta(\mathbf{X})$, another group of five variables make a small contribution to the variance, and the remaining five have very little effect. In the prior mean function we set $h(\mathbf{x}) = 1$, and we evaluate $\eta(\mathbf{x})$ at 250 design points. These design points are chosen to make the function $\int_\mathcal{X} c^*(\mathbf{x}, \mathbf{x}) dG(\mathbf{x})$ small, conditional on guessed values of the roughness parameters in B. By doing so we are attempting to make the posterior variance of $\eta(x)$ small, over the range of values of $\mathbf{x}$ of interest as defined by $G(\mathbf{x})$. Given the 250 runs, we find the posterior mode of B, and conditional on this estimate $Var(Y)$ and $Var_{X_i}\{E(Y|X_i)\}$ by their posterior expectations for each of the fifteen input variables. These posterior expectations can be computed analytically for our choice of $h(\mathbf{x})$ and input distribution G, conditional on B. The true values of these terms can also be determined analytically. The true and estimated main effects are displayed in figure 10.2. Again the errors in the estimated values are small, so in terms of contribution to the output variance, we have successfully identified the influence of each input variable. As 28% of the variance is unaccounted for by the main effect terms, this indicated additional variance due to interactions between inputs.

5. Conclusions

We have discussed various statistical issues related to the use of deterministic computer models. We have presented means of exploring output uncertainty that results from uncertainty in the model inputs. Our methods are especially suited to models that are computationally expensive, so that the number of runs of the model required is an important issue.

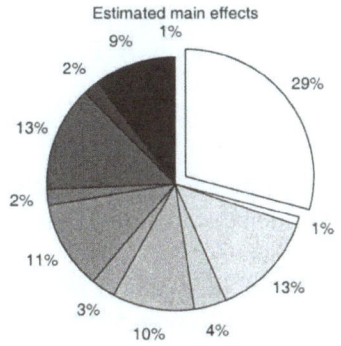 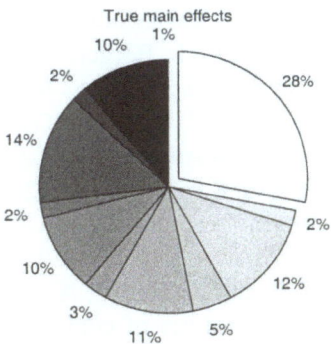

Figure 10.2. True and estimated main effects. The five variables with the smallest main effects have been grouped together and are shown as the top segement. The detached segment shows the variance remaining after all the main effects have been summed.

References

Clarke, R. H. (1979). The first report of a working group on atmospheric dispersion: a model for short and medium range dispersion of radionuclides released to the atmosphere. Technical Report NRPB-R91, National Radiological Protection Board.

Cox, D. C. (1982). An analytical method for uncertainty analysis of nonlinear output functions, with applications to fault-tree analysis. *IEEE Trans. Reliab*, R-31:265–268.

Craig, P. S., Goldstein, M., Rougier, J. C., and Seheult, A. H. (2001). Bayesian forecasting for complex systems using computer simulators. *J. Am. Statist. Assoc.*, 96:717–729.

Currin, C., Mitchell, T. J., Morris, M., and Ylvisaker, D. (1991). Bayesian prediction of deterministic functions with applications to the design and analysis of computer experiments. *J. Am. Statist. Assoc.*, 86:953–963.

Dunning, D. E., Schwarz, J. R., and Schwarz, G. (1981). Variability of human thyroid characteristics and estimates of dose from ^{131}I. *Health Physics*, 40.

Haylock, R. (1997). *Bayesian Inference about Outputs of Computationally Expensive Algorithms with Uncertainty on the Inputs*. PhD thesis, Department of Mathematics, University of Nottingham.

Haylock, R. G. and O'Hagan, A. (1996). On inference for outputs of computationally expensive algorithms with uncertainty on the inputs. In Bernardo, J. M., Berger, J. O., Dawid, A. P., and Smith, A. F. M., editors, *Bayesian Statistics 5*, pages 629–637. Oxford University Press.

Kennedy, M. C. and O'Hagan, A. (2001). Bayesian callibration of complex computer models (with discussion). *J. Roy. Statist. Soc. Ser. B*, 63:425–464.

Neal, R. (1999). Regression and classification using gaussian process priors. In Bernardo, J. M., Berger, J. O., Dawid, A. P., and Smith, A. F. M., editors, *Bayesian Statistics 6*, pages 69–95. Oxford University Press.

Oakley, J. E. (2002). Eliciting Gaussian process priors for complex computer codes. *The Statistician*, 51:81–97.

Oakley, J. E. and O'Hagan, A. (2000). Bayesian inference for the uncertainty distribution. Technical Report 497/00, Department of Probability and Statistics, University of Sheffield. To appear in *Biometrika*.

Oakley, J. E. and O'Hagan, A. (2002). A Bayesian approach to probabilistic sensitivity analysis of complex models. Technical Report 524/02, Department of Probability and Statistics, University of Sheffield.

O'Hagan, A. (1994). *Kendall's Advanced Theory of Statistics, Volume 2B, Bayesian Inference*. Edward Arnold, London.

Sacks, J., Welch, W. J., Mitchell, T. J., and Wynn, H. P. (1989). Design and analysis of computer experiments. *Statist. Sci.*, 4:409–435.

Saltelli, A., Chan, K., and Scott, M., editors (2000). *Sensitivity Analysis*. Wiley, New York.

Chapter 11

B

T. Poggio, S. Mukherjee, R. Rifkin, A. Raklin
McGovern Institute and Center for Biological and Computational Learning
Center for Genome Research, Whitehead Institute
Massachusetts Institute of Technology
Cambridge, MA 02139, USA
tp@ai.mit.edu, {sayan, rif, rakhlin}@mit.edu

A. Verri
INFM - DISI, Universita di Genova, Genova, Italy
verri@disi.unige.it

Abstract   In this chapter we summarize density properties of Reproducing Kernel Hilbert Spaces induced by different classes of kernels. They are important to characterize the "power" of the associated hypothesis spaces. In the process we characterize the role of b, which is the constant in the standard form of the solution provided by the Support Vector Machine technique $f(\mathbf{x}) = \sum_{i=1}^{\ell} \alpha_i K(\mathbf{x}, \mathbf{x}_i) + b$, which is a special case of Regularization Machines.

Keywords: RKHS, regularization, density

Introduction

Support Vector Machines (SVMs), originally introduced by Vapnik [Vapnik, 1995] in the context of Statistical Learning Theory, can be shown to be a special case of a regularization approach [Tikhonov and Arsenin, 1977] to the ill-posed problem of regression or classification from sparse and finite data [Evgeniou et al, 2000, Wahba, 1990]. The derivation of [Evgeniou et al, 2000, Girosi et al, 1990] makes it clear that SVMs and a large body of different learning and approximation techniques, known as Regularization Networks (RNs) [Girosi et al, 1995], can be obtained from the same general principles. Note that in the past [Evgeniou et al, 2000, Girosi et al, 1995] we have used the term RN to indicate the networks arising from regularization techniques, mainly involving the (classical) quadratic loss function. In this paper we use the term RN exclusively for the subset of regularization machines (RM) techniques that involve a quadratic loss function and call Regularization Machines the broader set of techniques (see equation 1.1 later) that includes SVMs and RNs as special cases.

The aim of this chapter is to characterize properties of a Reproducing Kernel Hilbert Space (RKHS) induced by positive definite and conditionally positive definite kernels. In the process, we will discuss the role of b, the constant term in the solution to the learning problem obtained in the standard derivation of SVMs due to Vapnik [Vapnik, 1995] and found also in some RNs. The issue is relevant a) for the theoretical connection between SVMs, RNs, and other techniques (see for instance [Girosi, 1998]) and b) for developing efficient algorithms for SVMs.

Throughout the chapter we assume some familiarity with both SVMs and RNs as presented in [Evgeniou et al, 2000].

1. Motivation

Given ℓ training pairs $\{(\mathbf{x}_1, y_1), ..., (\mathbf{x}_\ell, y_\ell)\}$ a regularized solution – that we call a *Regularization Machine* – to the learning problem is found by minimizing the following functional for fixed λ

$$I[f] = \frac{1}{\ell} \sum_{i=1}^{\ell} V(y_i, f(\mathbf{x}_i)) + \lambda \|f\|_K^2 \qquad (11.1)$$

corresponding to the minimization of the empirical loss – the first term – under capacity control – the second term. The choice of the loss function V determines the learning scheme. In classical (quadratic) Regularization Networks: $V(y_i, f(\mathbf{x}_i)) = (y_i - f(\mathbf{x}_i))^2$, in SVM Classification: $V(y_i, f(\mathbf{x}_i)) = |1 - y_i f(\mathbf{x}_i)|_+$ where $|x|_+ = \max\{x, 0\}$, and in SVM Regression: $V(y_i, f(\mathbf{x}_i)) = |y_i - f(\mathbf{x}_i)|_\epsilon$ where $|x|_\epsilon \equiv \max\{0, |x| - \epsilon\}$ is called ϵ-insensitive loss.

The term $\|f\|_K$ is the norm of the function f in the RKHS induced by a positive definite kernel K [Aronszajn, 1950, Wahba, 1990]. It has been known for some time (see [Girosi and Poggio, 1990]) that the minimizer of $I[f]$ for rather general $V(\cdot)$ and in particular for RNs [Wahba, 1990, Girosi et al, 1995] and SVMs [Vapnik, 1998, Girosi, 1998, Evgeniou et al, 2000, Lin et al, 2000] belongs to the RKHS induced by K and can be written as

$$f(\mathbf{x}) = \sum_{i=1}^{\ell} \alpha_i K(\mathbf{x}, \mathbf{x}_i)$$

for some coefficients α_i which depend on the ℓ examples and on λ. In the original formulation of SVMs due to Vapnik, the minimizer is actually written

b

as

$$f(\mathbf{x}) = \sum_{i=1}^{\ell} \alpha_i K(\mathbf{x}, \mathbf{x}_i) + b \qquad (11.2)$$

with $-\frac{1}{2\lambda\ell} \leq \alpha_i \leq \frac{1}{2\lambda\ell}$ and

$$\sum_{i=1}^{\ell} \alpha_i = 0. \qquad (11.3)$$

The offset parameter b is estimated (like the coefficients α_i) from the ℓ examples. From the dual formulation of the optimization problem of SVMs (see for example [Vapnik, 1995]), it can be seen that the equality constraint (11.3) is induced[1] by the form of the solution assumed in equation (11.2).

The case of standard RNs (quadratic V) is well known (see [Wahba, 1990] or [Girosi and Poggio, 1990]). If the kernel is a positive definite function (i.e., in the case of Gaussian Radial Basis Functions) one chooses the solution in the linear span of the RKHS written as $f(\mathbf{x}) = \sum_{i=1}^{\ell} \alpha_i K(\mathbf{x}, \mathbf{x}_i)$ without any constraint on the α_i. If the kernel is a conditionally positive definite function of order 1 (i.e., in the case of piecewise linear splines) a constant term, b, is added to the solution which becomes $f(\mathbf{x}) = \sum_{i=1}^{\ell} \alpha_i K(\mathbf{x}, \mathbf{x}_i) + b$, subject to the same constraint (11.3). As shown in [Evgeniou et al, 2000] the solution $f(\mathbf{x}) = \sum_{i=1}^{\ell} \alpha_i K(\mathbf{x}, \mathbf{x}_i) + b$ with the constant b (and the equality constraint in (11.3)) can be used also in the case of a positive definite kernel K.[2]

This chapter is devoted to answering the following questions: When should b be used? Is there a choice of using or not using b? What does the choice mean? Are the answers different for RNs and SVMs? And most importantly: what is the relation between properties of the kernel K and approximation properties of the corresponding RKHS?

2. Analysis
2.1 Definitions

First we define (conditional) positive definiteness of kernels. More properties of shift invariant kernels, $K(x, y) = K(x - y)$, can be found in [Schoenberg, 1998, Berg et al, 1984, Micchelli, 1986].

---

[1]This property of b was not discussed in [Vapnik, 1995].
[2]This choice – rather than the usual one without b – effectively corresponds to a different (semi)norm and to a different RKHS. Equation (4.10) in [Evgeniou et al, 2000] is incorrect and should be replaced with $(K' + \ell\lambda I)\alpha + \mathbf{1}b = (K + \ell\lambda I)\alpha + \mathbf{1}b = \mathbf{y}$. The key equation (4.12) and the conclusions are however correct. Notice that an unfortunate error (noticed by Steve Smale) propagates throught three other equations in [Evgeniou et al, 2000]: λ should be replaced by $\lambda\ell$ in Equations (4.3, 4.6, 4.19).

Let X be some set, for example a subset of $\mathbb{R}^d$ or $\mathbb{R}^d$ itself. A *kernel* is a symmetric function $K : X \times X \to \mathbb{R}$.

DEFINITION 11.1

A kernel $K(\mathbf{t}, \mathbf{s})$ is positive definite (pd) *if $\sum_{i,j=1}^{n} c_i c_j K(\mathbf{t}_i, \mathbf{t}_j) \geq 0$ for any $n \in \mathbb{N}$ and choice of $\mathbf{t}_1, ..., \mathbf{t}_n \in X$ and $c_1, ..., c_n \in \mathbb{R}$.*

An equivalent definition could be given in terms of positive semidefiniteness of the matrix $K_{ij} = K(\mathbf{t}_i, \mathbf{t}_j)$. A *pd* kernel is *strictly positive definite* if for any distinct vectors $\mathbf{t}_1, ..., \mathbf{t}_n \in X$ the above inequality holds strictly when the c_i are not all zero (in that case the matrix K_{ij} is positive definite and not just positive semidefinite).

DEFINITION 11.2

A kernel $K(\mathbf{t}, \mathbf{s})$ is conditionally positive definite (cpd) *of order 1 if*

$$\sum_{i,j=1}^{n} c_i c_j K(\mathbf{t}_i, \mathbf{t}_j) \geq 0$$

for any $n \in \mathbb{N}$ and choice of $\mathbf{t}_1, ..., \mathbf{t}_n \in X$ and $c_1, ..., c_n \in \mathbb{R}$ subject to the constraint $\sum_{i=1}^{n} c_i = 0$. It is strictly conditionally positive definite if $\sum_{i,j=1}^{n} c_i c_j K(\mathbf{t}_i, \mathbf{t}_j) > 0$.

2.2 Integral operators

We consider the integral operator L_K on $L_2(X, \nu)$ defined by

$$\int_X K(\mathbf{x}, \mathbf{x}') f(\mathbf{x}') d\nu(\mathbf{x}') = g(\mathbf{x}) \qquad (11.4)$$

where X is a compact subset of $\mathbb{R}^n$ and ν a Borel measure. We assume K to be continuous. Thus the integral operator is compact [Cucker and Smale, 2001]. Note that K pd (definition 11.1) is equivalent [Mercer, 1909] to L_K positive, that is

$$\int_X K(\mathbf{t}, \mathbf{s}) f(\mathbf{t}) f(\mathbf{s}) d\nu(\mathbf{t}) d\nu(\mathbf{s}) \geq 0 \qquad (11.5)$$

for all $f \in L_2(X, \nu)$.

2.3 Mercer's theorem

The key tool in our analysis is the result published by Mercer in 1909 [Mercer, 1909, Courant and Hilbert, 1962].

THEOREM 2.1 *A symmetric, pd kernel $K : X \times X \to \mathbb{R}$, with X a compact subset of $\mathbb{R}^n$ has the expansion*

$$K(\mathbf{x}, \mathbf{x}') = \sum_{q=1}^{\infty} \mu_q \phi_q(\mathbf{x}) \phi_q(\mathbf{x}') \tag{11.6}$$

where the convergence is in $L_2(X, \nu)$. If the measure ν on X is non-degenerate in the sense that open sets have positive measure everywhere, then the convergence is absolute and uniform and the $\phi(\mathbf{x})$ are continuous on X (see [Cucker and Smale, 2001] and Smale, pers. comm.). The ϕ_q are the orthonormal eigenfunctions of the integral equation

$$\int_X K(\mathbf{x}, \mathbf{x}') \phi(\mathbf{x}) d\nu(\mathbf{x}) = \mu \, \phi(\mathbf{x}'). \tag{11.7}$$

Using Mercer's theorem we distinguish three cases, depending on the properties of the kernel.

1. *Strictly positive case:* The kernel is *strictly pd* and all eigenvalues (there is an infinite number of eigenvalues) of the integral operator L_K are strictly positive.

2. *Degenerate case:* The kernel $K(\mathbf{t}, \mathbf{s})$ is positive definite but only a finite number of eigenvalues of the integral operator L_K are strictly positive, the rest being zero (see [Courant and Hilbert, 1962]).

3. *Conditionally strictly positive case:* The kernel $K(\mathbf{t}, \mathbf{s})$ is *conditionally positive definite* and all the eigenvalues of the integral operator L_K are positive with only a finite number being non-positive. Notice that for cpd kernels, the kernel K can be made into a positive definite kernel K' by subtracting the terms $\mu_q \phi_q(\mathbf{x}) \phi_q(\mathbf{x}')$ belonging to negative eigenvalues [Courant and Hilbert, 1962].

2.4 Reproducing Kernel Hilbert Spaces

The RKHS induced by K is equivalent (see [Cucker and Smale, 2001]) to the Hilbert space of the functions spanned by $\Phi = \{\phi_1(\mathbf{x}), ...\}$,

$$f(\mathbf{x}) = \sum_{q=1}^{\infty} c_q \phi_q(\mathbf{x}),$$

equipped with the scalar product $< f, g > = \sum_{q=1}^{\infty} \frac{c_q d_q}{\mu_q}$ where

$$f(\mathbf{x}) = \sum_{q=1}^{\infty} c_q \phi_q(\mathbf{x}), \qquad g(\mathbf{x}) = \sum_{q=1}^{\infty} d_q \phi_q(\mathbf{x}),$$

and finite norm in the RKHS $\|f\|_K^2 = \sum_{q=1}^{\infty} c_q^2/\mu_q$, where the sums for the norm and scalar products are over terms with nonzero μ_q. Note that it is possible to prove directly that the RKHS is independent of the measure ν (assumed positive everywhere), as observed by Smale and Cucker (though the $\phi_q(\mathbf{x})$ and the μ_q in equation (11.7) are not).

2.5 Density of a RKHS

We characterize density properties of a RKHS. In particular, we ask under which condition is a RKHS dense in $L_2(X, \nu)$.[3] The answer below was developed starting from separate observations by Zhou [personal communication], Girosi [personal communication] and Smale [personal communication]. (In the following we assume ν to be the Lebesgue measure.) The following statements follow for the three cases above:

1. In the *strictly positive case* the RKHS is infinite dimensional and dense in $L_2(X, \nu)$. The proof depends on the observation (see for instance [Kolmogorov, A.N. and Fonim, S.V., 1957], Theorem 3 following the Riesz-Fisher theorem) that if the RKHS induced by K were not dense in L_2 there would exist a non trivial function f orthogonal to the RKHS; but this would imply that L_K would have zero eigenvalues, contrary to the hypothesis.

2. In the *degenerate case* the RKHS is finite dimensional and not dense in $L_2(X, \nu)$; the null space of the operator L_K is infinite dimensional.

3. In the *conditionally strictly positive case* the RKHS associated with K' is infinite dimensional and the null space of the operator L_K is finite dimensional. The RKHS is not dense in $L_2(X, \nu)$ but when completed with a finite number of polynomials of appropriate degree can be made to be dense in $L_2(X, \nu)$.

2.6 Regularization Networks (including SVMs) for Regression :

In regression, given sparse data it is natural and desirable to be able to approximate the unknown function under the most general conditions, such as all functions in $L_2(X, \nu)$. From this perspective we look at possible solutions of (11.1) for the three cases above.

---

[3]Density in $C(X)$ (in the sup norm) is a trickier issue that has been answered very recently by Zhou (in preparation). It is guaranteed for radial kernels if density in $L_2(X, \nu)$ holds.

1. *Strictly positive case*: In this case the solution

$$f(\mathbf{x}) = \sum_{i=1}^{\ell} \alpha_i K(\mathbf{x}, \mathbf{x}_i) \qquad (11.8)$$

is dense in $L_2(X, \nu)$: b is not needed. Note that this is a different kind of SVM from the one originally proposed by Vapnik, even if Vapnik's loss functions are used. However, the solution with b that Vapnik originally proposed, is also valid since a positive definite kernel K is also *cpd*. It is easy to check (see the following cpd case) that using the solution with b is equivalent to using the cpd kernel $K'(\mathbf{x}, \mathbf{y}) = K - \mu_1 \phi_1(\mathbf{x}) \phi_1(\mathbf{y})$ in the stabilizer term of equation (11.1) with a solution $f(\mathbf{x}) = \sum_{i=1}^{\ell} \alpha'_i K'(\mathbf{x}, \mathbf{x}_i) + b$. Somewhat surprisingly, it follows that

$$f(\mathbf{x}) = \sum_{i=1}^{\ell} \alpha'_i K'(\mathbf{x}, \mathbf{x}_i) + b = \sum_{i=1}^{\ell} \alpha'_i K(\mathbf{x}, \mathbf{x}_i) + b. \qquad (11.9)$$

Thus, in this case, both solutions (11.8) and (11.9) are dense in $L_2(X, \nu)$. Notice that they correspond, respectively, to the minimizers of the following functionals, each one using a different prior on the function space

$$I[f] = \frac{1}{\ell} \sum_{i=1}^{\ell} V(y_i, f(\mathbf{x}_i)) + \lambda \|f\|_K^2$$

$$I[f] = \frac{1}{\ell} \sum_{i=1}^{\ell} V(y_i, f(\mathbf{x}_i)) + \lambda \|f\|_{K'}^2.$$

In the RN case with quadratic $V(\cdot)$ the minimization of the two different $I[f]$ yields the linear equations (see [Evgeniou et al, 2000]): $(K + \ell \lambda I)\alpha = \mathbf{y}$ and $(K + \ell \lambda I)\alpha' + \mathbf{1}b = \mathbf{y}$ subject to $\mathbf{1}\alpha = 0$. Thus in the standard RN case it is possible to compute one solution from the other since $\alpha - \alpha' = (K + \ell \lambda I)^{-1} \mathbf{1}b$. In the SVM case the two different solutions correspond to the minima of two different QP problems and the relation between α and (α', b) cannot be given in closed form.

2. *Degenerate case*: In this case the regularization solution is not dense in $L_2(X)$ with or without the addition of a polynomial of finite degree. In other words, with a finite dimensional kernel it is in general impossible to approximate arbitrarily well a continuous function on a bounded interval. This is the case for polynomial kernels of the form

$K(\mathbf{x}, \mathbf{y}) = (\mathbf{x} \cdot \mathbf{y} + 1)^n$ often used in SVMs. The use of b here is therefore even more arbitrary (or more dependent on prior knowledge about the specific problem), since it does not restore density.

3 *Conditionally strictly positive case*: In this case [4] the solution

$$f(\mathbf{x}) = \sum_{i=1}^{\ell} \alpha_i K(\mathbf{x}, \mathbf{x}_i) + b = \sum_{i=1}^{\ell} \alpha_i K_{pd}(\mathbf{x}, \mathbf{x}_i) + b$$

is dense in $L_2(X, \nu)$ when the b term is included. We define the *pd* kernel $K_{pd} = K - \mu_0 \phi_1 \phi_1$ where $-\mu_0 \phi_1 \phi_1$ is a positive constant (we assume here, since we are dealing with cpd kernels of degree 1, that $\phi_1(\mathbf{x})\phi_1(\mathbf{x}')$ corresponds to a constant). The stabilizer term in $I[f]$ is then formally interpreted (compare [Aronszajn, 1950]) as $\|f\|_K^2 = \sum_{q=2}^{\infty} \frac{c_q^2}{\mu_q}$, that is as $\|f\|_K^2 = \|f\|_{K_{pd}}^2 = \|P_{K_{pd}} f\|_{K_{pd}}^2$, where $P_{K_{pd}}$ projects f into the RKHS induced by K_{pd} (see [Wahba, 1990]). To obtain solutions that are dense in $L_2(X, \nu)$ we consider solutions of the form $f(\mathbf{x}) = \sum_{q=2}^{\infty} c_q \phi_q(\mathbf{x}) + b$. These are functions in the RKHS induced by the *pd* kernel K_{pd} completed with the constants (which are not in the RKHS). Taking derivatives of $I[f]$ with respect to the coefficients c_q and b and setting them equal to zero (following [Girosi, 1998]) we get $c_q = \mu_q \sum_{i=1}^{\ell} \alpha_i \phi_q(\mathbf{x}_i)$ where $\alpha_i = \frac{1}{\lambda} V'(y_i - f(\mathbf{x}_i))$, subject to the constraint

$$\sum_{i=1}^{\ell} \alpha_i = 0. \qquad (11.10)$$

Therefore the minimizer of $I[f]$ is

$$\begin{aligned} f(\mathbf{x}) &= \sum_{q=2}^{\infty} c_q \phi_q(\mathbf{x}) + b = \sum_{i=1}^{\ell} \alpha_i \sum_{q=2}^{\infty} \mu_q \phi_q(\mathbf{x}_i) \phi_q(\mathbf{x}) + b \\ &= \sum_{i=1}^{\ell} \alpha_i K_{pd}(\mathbf{x}, \mathbf{x}_i) + b = \sum_{i=1}^{\ell} \alpha_i K(\mathbf{x}, \mathbf{x}_i) + b \end{aligned}$$

where we have used expansion 11.6 and, in the last step, constraint 11.10. By interpreting $V'(\cdot)$ in a generalized sense, the proof is valid

---

[4] For simplicity we consider in the following cpd kernels of order 1 only.

for a broad class of $V(\cdot)$ (see [Girosi et al, 1990]) and for the non-differentiable $V(\cdot)$ used by SVM regression and classification see appendix B.2 of [Girosi, 1998]. Thus cpd kernels can be used not only for standard RNs but also for SVMs[5]. In both cases the term b is needed in the solution in order to approximate functions in $L_2(X,\nu)$.

Notice that the density of the solutions of regularization in $L_2(X,\nu)$ implies that RKHSs with a positive definite infinite dimensional kernel (or cpd if completed with polynomials) correspond to quite large hypothesis space. The property thus provides a general justification for the use of RKHS as hypothesis spaces for learning problems.

3. Regression and Classification: b or not b?

Let us consider here only the *strictly positive case* for K. In regression the general solution does not have a b term. However, it is possible to use a positive definite K *and* the constant b. The latter choice is effectively the choice of a different kernel and a different feature space relative to the initial K used in the standard solution without b: the constant feature "disappears" from the RKHS norm and therefore is not "penalized". This choice may be reasonable but only when specific prior information is available about the problem. For instance, there may be regression problems in which shifts of f by a constant should not be penalized. This is especially true for the binary classification framework originally considered by Vapnik. Only the sign of the function f found by the SVM is used for classification. A constant b plays therefore the role of a threshold; using a solution of the form $f(\mathbf{x}) = \sum_{i=1}^{\ell} \alpha_i K(\mathbf{x}, \mathbf{x}_i) + b$, without penalizing b in the stabilizer, corresponds to the reasonable assumption that there is no privileged value – such as 0 – for the classification threshold.

Acknowledgments

We wish to thank Federico Girosi for many discussions and a legacy of insights, Steve Smale for very useful suggestions and Ding-Xuan Zhou for quickly providing the answer to our question on density of RKHS and for much help afterwards.

References

[Aronszajn, 1950] N. Aronszajn "Theory of reproducing kernels," *Trans. Amer. Math. Soc.*, vol. 686, pp. 337-404, 1950.

---

[5]This extension was well known for regularization since at least a decade; for SVMs see [Girosi, 1998]

[Berg et al, 1984] C. Berg, J.P.R. Christensen, and P. Ressel *Harmonic Analysis on Semigroups: Theory of Positive Definite and Related Functions* Springer, 1984.

[Courant and Hilbert, 1962] R. Courant and D. Hilbert *Methods of Mathematical Physics, Vol 2* Interscience, 1962.

[Cucker and Smale, 2001] F. Cucker and S. Smale On the Mathematical Foundations of Learning, *Bull. Amer. Math. Soc.*, vol. 39, pp. 1–49, 2002.

[Evgeniou et al, 2000] T. Evgeniou, M. Pontil, and T. Poggio "Regularization Networks and Support Vector Machines," *Advances in Computational Mathematics*, vol. 13, pp. 1–50, 2000.

[Freiss et al, 1998] T. Friess, N. Cristianini, and C. Campbell "The Kernel-Adatron: A fast and simple learning procedure for Support Vector Machines," *Proceedings of the 15th International Conference in Machine Learning*, pages 188-196, 1998.

[Girosi, 1998] F. Girosi "An equivalence between sparse approximation and support vector machines," *Neural Computation*, vol. 10, pp. 1455–1480, 1998.

[Girosi et al, 1990] F. Girosi and T. Poggio and B. Caprile Extensions of a Theory of Networks for Approximation and Learning: outliers and Negative Examples, Advances in Neural information processings systems 3, R. Lippmann and J. Moody and D. Touretzky, Morgan Kaufmann, 1991, San Mateo, CA.

[Girosi and Poggio, 1990] F. Girosi and T. Poggio "Networks and the Best Approximation Property," *Biological Cybernetics*, vol. 63, pp. 169–176, 1990.

[Girosi et al, 1995] F. Girosi, M. Jones, and T. Poggio "Regularization theory and neural network architectures," *Neural Computation,* vol. 7, pp. 219–269, 1995.

[Kolmogorov, A.N. and Fonim, S.V., 1957] A.N. Kolmogorov and S.V. Fonim *Elements of the Theory of Functions and Functional Analysis* Dover, 1957

[Lin et al, 2000] Y. Lin, Y. Lee, and G. Wahba *Support Vector Machines for Classification in Nonstandard Situations* TR 1016, March 2000. To appear, Machine Learning

[Mercer, 1909] J. Mercer "Functions of positive and negative type and their connection with the theory of integral equations," *Phil.los. Trans. Roy. Soc. London Ser. A*, vol. 209, pp. 415–446, 1909.

[Micchelli, 1986] C.A. Micchelli "Interpolation of scattered data: distance matrices and conditionally positive definite functions," *Constructive Approximation*, vol. 2, pp. 11–22, 1986.

[Poggio et al., 2001] T. Poggio, S. Mukherjee, R. Rifkin, A. Rakhlin and A. Verri " b," CBCL Paper 198/AI Memo 2001-011, Massachusetts Institute of Technology, Cambridge, MA, July 2001

[Schoenberg, 1998] I.J. Schoenberg "Metric spaces and completely monotone functions" *Ann. of Math.*,vol. 39, pp.811–841, 1938.

[Tikhonov and Arsenin, 1977] A. N. Tikhonov and V. Y. Arsenin Solutions of Ill-posed Problems W. H. Winston, 1977

[Vapnik, 1995] V.N. Vapnik The Nature of Statistical Learning Theory Springer, 1995.

[Vapnik, 1998] V.N. Vapnik Statistical Learning Theory Wiley, 1998.

[Wahba, 1990] G. Wahba *Spline models for observational data* SIAM, 1990.

Chapter 12

AFFINE ARITHMETIC AND BERNSTEIN HULL METHODS FOR ALGEBRAIC CURVE DRAWING

Huahao Shou, Ralph Martin
Department of Computer Science, Cardiff University, Cardiff, United Kingdom
{h.shou,ralph.martin}@cs.cf.ac.uk

Guojin Wang
Department of Mathematics, Zhejiang University, Hangzhou, China
wgj@math.zju.edu.cn

Irina Voiculescu
Computing Laboratory, Oxford University, Oxford, UK
irina@comlab.ox.ac.uk

Adrian Bowyer
Department of Mechanical Engineering, Bath University, Bath, UK
A.Bowyer@bath.ac.uk

Abstract   We compare approaches to the location of the algebraic curve $f(x,y) = 0$ in a rectangular region of the plane, based on recursive use of conservative estimates of the range of the function over a rectangle. Previous work showed that performing interval arithmetic in the Bernstein basis is more accurate than using the power basis, and that affine arithmetic in the power basis is better than using interval arithmetic in the Bernstein basis. This paper shows that using affine arithmetic with the Bernstein basis gives no advantage over affine arithmetic with the power basis. It also considers the Bernstein coefficient method based on the convex hull property, which has similar performance to affine arithmetic.

Keywords: Interval arithmetic, affine arithmetic, Bernstein hull, curve drawing.

1. Introduction

Solving $f(x,y) = 0$ in a rectangular area $[\underline{x}, \overline{x}] \times [\underline{y}, \overline{y}]$, where $f(x,y)$ is a polynomial, is a problem with many practical applications in CAD and computer graphics. One such example is drawing the algebraic curve represented by $f(x,y) = 0$; other applications include surface-surface intersection and silhouette edge detection of a parametric surface [12, 15, 17].

Let C be an algebraic curve defined by the equation $f(x,y) = 0$. A simple and general technique for computing an approximation of C on a rectangular region Ω as described in [9] is:

(1) decompose Ω into small cells, typically on a rectangular grid;
(2) identify which cells intersect C;
(3) approximate C within each intersecting cell (e.g. fill the cell if it is a screen pixel).

Finding the intersecting cells is usually the most expensive step in this method. The simplest approach is to test all cells, but this is computationally very expensive.

Several methods exist for finding the cells intersecting a curve C without visiting all the cells in a cellular decomposition of a region Ω. Continuation methods [4] sample the curve only in the immediate neighbourhood of known intersecting cells. They start from one or more seed cells known to contain the curve, and follow it into adjacent cells. The fundamental difficulty of this approach is finding a complete set of initial seed cells intersecting *every* connected component of C in Ω.

Large portions of Ω can be discarded quickly and reliably if the absence of C in a particular region can be proved. Hierarchical decomposition methods [6, 15, 16, 17] rely on such a test to explore Ω recursively, starting with Ω itself as the initial cell. If a cell is proved to be empty, it is ignored; otherwise, it is subdivided into smaller cells, which are then explored recursively, until the cells are small enough to approximate C. Range analysis is used to test if the curve passes through a cell. Range analysis methods output a range of values guaranteed to *include* the values the function takes over a given range of values for x and y. Interval arithmetic (IA) [11] provides a natural tool for range analysis [13]. Hierarchical decomposition methods based on IA have been widely used in computer graphics applications [6, 15, 17].

The main weakness of IA is that it tends to be too conservative [5, 8, 9], i.e. the range for the function output by IA is much wider than the actual range of values the function takes over a given interval. To solve this problem, Comba and Stolfi [5] proposed a new model for numerical computation, called affine arithmetic (AA). Both IA and AA can be used to manipulate imprecise values and to evaluate functions over intervals. Both can also keep track of truncation and round-off errors. However, in contrast to IA, AA maintains dependencies between the sources of error arising from different variables, and thus manages to compute significantly tighter error bounds. AA has been used as a replacement for IA in various computer graphics applications, such as ray tracing, intersection testing, enumeration of implicit curves and surfaces, and sampling for procedural shaders [5, 8, 9, 10]. As AA computes tighter intervals than IA, it is possible to draw algebraic curves using the former more efficiently and with higher quality than using the latter [9, 19].

Because of the way arithmetic operators work in IA, the basis used to express the polynomial $f(x,y)$ affects the range for the function output by IA evaluation. IA using the Bernstein basis is more accurate than IA using the power basis. Even more, AA using the power basis is better than IA using the Bernstein basis [18]. The same paper [18] asked whether AA using the Bernstein basis would be even better. The current paper shows that AA in the Bernstein basis has no advantage over AA in the power form.

Range analysis may also be performed using the Bernstein coefficient (BC) method based on the Bernstein convex hull property [13]. If a polynomial is written in the Bernstein basis, its range is bounded by the values of the minimum and maximum Bernstein coefficients. We compare the performance and efficiency of the AA method and the BC method; we also show results of using IA methods using the power basis and Bernstein basis.

2. Algebraic curve drawing algorithm

The basic strategy [19] for drawing an algebraic curve $f(x,y) = 0$ in a given rectangular interval $[\underline{x}, \overline{x}] \times [\underline{y}, \overline{y}]$ is to evaluate $f(x,y)$ over the desired interval using some range analysis evaluation method giving a range $\mathbf{F} = [\underline{F}, \overline{F}]$. If the resulting interval does not contain 0, the curve *cannot* be present. If it does contain 0, the region may or may not contain the mathematical curve, so the interval is divided horizontally and vertically at its mid-point, and the pieces are considered in turn. The process stops when an interval consisting of a single pixel is left. In such a case we fill the pixel anyway, even though we are still unsure whether the curve is actually present. This may result in a "fat" curve if the test is too conservative, i.e. pixels may be filled which do not actually contain the curve.

3. Interval arithmetic method

Traditional IA methods are used as a reference for comparison in this paper. An interval $\mathbf{x} = [a, b]$ is a set of real numbers defined by $[a, b] = \{x | a \leq x \leq b\}$. Rules used to perform arithmetic on intervals may be found in [11].

The natural interval extension $\mathbf{f}(\mathbf{x}, \mathbf{y})$ of a bivariate polynomial $f(x, y)$ is obtained by replacing each occurrence of x and y in $f(x, y)$ by intervals $\mathbf{x}$ and y, and evaluating the resulting interval expression using the above definitions. The result is itself an interval.

As already noted, when using IA, the basis used for the polynomial expression can affect the result [3]. Suitably rearranging the function can give tighter bounds on the result, although the result will be still conservative, not exact.

3.1   IA using the power basis

Here we describe how to use IA to evaluate a polynomial in two variables written in the power basis. Let

$$\mathbf{f}(\mathbf{x},\mathbf{y}) = \sum_{i=0}^{n}\sum_{j=0}^{m} A_{ij} x^i y^j, \quad (x,y) \in \Omega = [\underline{x},\overline{x}] \times [\underline{y},\overline{y}].$$

It is helpful to rewrite it in matrix representation: $\mathbf{f}(\mathbf{x},\mathbf{y}) = XAY$, where $X = (1, x, ..., x^n), Y = (1, y, ..., y^m)^T$. X and Y are first computed using IA rules, and then the matrix product is found. The matrix representation of the polynomial power form plays a crucial role in the matrix AA evaluation method [14].

3.2   IA using the Bernstein basis

The Bernstein basis is widely used for the generation of Bezier, B-spline and NURBS curves and surfaces [2]. Bowyer, Berchtold, and Voiculescu [1, 2, 3, 18] have extensively considered the use of IA and multivariate Bernstein-form polynomials in geometric modelling. Conversion from the power form to the Bernstein form is discussed in [1, 2].

The Bernstein form often has many more terms than the power form. Furthermore, these contain repeated subexpressions of $x, (1-x), y$ and $(1-y)$. As repeated expressions can lead to excessive conservativeness in interval arithmetic, one might doubt the performance of the Bernstein form with interval arithmetic. Surprisingly, results show that the Bernstein form generally does better than the much simpler power form.

4.   Affine arithmetic method

As noted, the main weakness of IA is that it tends to be too conservative. In long computation chains, where the intervals computed by one stage are the inputs to the following stage, the relative accuracy of the computed intervals decreases exponentially. Unfortunately, long computation chains are not uncommon in geometric computing applications.

To address the error explosion problem in IA, Comba and Stolfi [5] proposed *affine arithmetic* (AA). Like IA, AA keeps track automatically of the round-off and truncation errors affecting each computed quantity. Unlike IA, however,

Algebraic Methods for Curve Drawing

AA keeps track of correlations between quantities. For example, in computing $x \times x$, IA treats each x as if it were a separate quantity, and would allow each x independently to be anywhere in its range. In contrast, AA "knows" they are the same variable—if the first x is near the lower end of its range, so is the second x. As a result, AA is able to provide much tighter intervals than IA, especially in long computation chains.

In AA, an uncertain quantity x is represented by an affine form $\hat{x}$ that is a first-degree polynomial in a set of noise symbols ε_i:

$$x = \hat{x} = x_0 + x_1\varepsilon_1 + \cdots + x_m\varepsilon_m = x_0 + \sum_{i=1}^{m} x_i\varepsilon_i.$$

Here the values of the noise symbols ε_i are unknown but are in the range $[-1, 1]$, and represent the uncertainty in x. Each corresponding coefficient x_i is a real number that determines the magnitude and sign of ε_i. Each ε_i stands for an independent source of uncertainty contributing to the total uncertainty in x. One may make m as large as necessary in order to represent all sources of uncertainty. (When converting an interval to AA form, we start with $m = 1$, but m may increase as nonlinear computations are performed, as explained later.) If the same noise symbol ε_i appears in two or more affine forms (e.g. in both $\hat{x}$ and $\hat{y}$), it indicates that some dependencies and correlations exist between the underlying quantities x and y.

An ordinary interval $[\underline{x}, \overline{x}]$ representing a quantity x may be written in affine form as [5, 8, 9]

$$\hat{x} = x_0 + x_1\varepsilon_x, \quad \text{where} \quad x_0 = (\underline{x} + \overline{x})/2, \quad x_1 = (\overline{x} - \underline{x})/2.$$

Conversely, given an affine form $\hat{x} = x_0 + x_1\varepsilon_1 + \cdots + x_m\varepsilon_m$, the range of possible values of the corresponding interval is

$$[\underline{x}, \overline{x}] = [x_0 - \xi, x_0 + \xi], \quad \text{where} \quad \xi = \sum_{i=1}^{m} |x_i|.$$

Various simple arithmetic operations are defined for AA in [5]. In AA, $\hat{x} - \hat{x} = 0$ and $(2\hat{x} + \hat{y}) - \hat{x} = \hat{x} + \hat{y}$, whereas IA produces wider intervals. The multiplication of two affine forms $\hat{x} \times \hat{y}$ produces a quadratic polynomial in the noise symbols ε_i. The quadratic term is replaced by a suitable coefficient multiplying a new noise symbol ε_k. This approximation may still produce a resulting interval up to four times as wide as the exact range of the quadratic term.

Unlike IA, which does not obey the distributive law, AA satisfies all commutative, associative and distributive laws. In this respect, there is no difference between AA and real arithmetic. Now, the various ways of expressing a polynomial function using different bases do nothing other than rearrange the terms. This does not affect the arithmetic of the polynomial, and hence does not affect the result of applying AA to an equivalent polynomial form. Therefore, when AA is involved, we only need to consider the power basis. (One may argue that in practice this is not quite true because of the non–commutativity, associativity and distributivity of computer real arithmetic. However, machine precision is negligible when compared with the length of the intervals used in solving the curve drawing problem.)

Since multiplication of affine forms may produce results much larger than the exact range, we use the matrix AA polynomial evaluation method proposed in [14], as below. This provides a better estimate for the range of a polynomial, using AA.

This works by first converting the interval forms $[\underline{x}, \overline{x}]$ and $[\underline{y}, \overline{y}]$ to affine forms: $\hat{x} = x_0 + x_1 \varepsilon_x$, $\hat{y} = y_0 + y_1 \varepsilon_y$. Then let

$$\hat{X} = (1, \varepsilon_x, ..., \varepsilon_x^n), \quad \hat{Y} = (1, \varepsilon_y, ..., \varepsilon_y^m)^T,$$

and define matrices B and C such that

$$B_{ij} = \begin{cases} \binom{j}{i} x_0^{j-i} x_1^i, & i \leq j \\ 0, & i > j \end{cases}, \quad i = 0, 1, ..., n; \quad j = 0, 1, ..., n,$$

$$C_{ij} = \begin{cases} 0, & i < j \\ \binom{i}{j} y_0^{i-j} y_1^j, & i \geq j \end{cases}, \quad i = 0, 1, ..., m; \quad j = 0, 1, ..., m.$$

Now, if we compute D from B and C, and the original coefficient matrix A using $D = BAC$, we get

$$f(\hat{x}, \hat{y}) = \hat{X} D \hat{Y} = \sum_{i=0}^{n} \sum_{j=0}^{m} D_{ij} \varepsilon_x^i \varepsilon_y^j$$

This is the exact affine form which we now wish to convert back to interval form $[\underline{F}, \overline{F}]$. The conversion procedure works as follows. If i is even and j is even, then $\varepsilon_x^i \varepsilon_y^j \in [0, 1]$, otherwise $\varepsilon_x^i \varepsilon_y^j \in [-1, 1]$. Thus:

$$\overline{F} = D_{00} + \sum_{j=1}^{m} \begin{cases} \max(0, D_{0j}), & \text{if } j \text{ is even} \\ |D_{0j}|, & \text{otherwise} \end{cases}$$

$$+ \sum_{i=1}^{n} \begin{cases} \max(0, D_{i0}), & \text{if } i \text{ is even} \\ |D_{i0}|, & \text{otherwise} \end{cases}$$

$$+ \sum_{i=1}^{n} \sum_{j=1}^{m} \begin{cases} \max(0, D_{ij}), & \text{if } i, j \text{ are both even} \\ |D_{ij}|, & \text{otherwise} \end{cases}$$

and similarly for $\underline{F}$.

5. Bernstein coefficient method

A different method for bounding a polynomial over an interval depends on the use of the Bernstein convex hull property [13]. It guarantees that the value of a polynomial in the Bernstein basis is bounded by the values of the minimum and maximum Bernstein coefficients.

To utilize the Bernstein convex hull property for evaluation of $f(x, y)$ in the region $[\underline{x}, \overline{x}] \times [\underline{y}, \overline{y}]$, we must first convert the range $[\underline{x}, \overline{x}] \times [\underline{y}, \overline{y}]$ to $[0, 1] \times [0, 1]$. This can be done by a change of variables to $\tilde{x}$ and $\tilde{y}$:

$$x = \underline{x} + (\overline{x} - \underline{x})\tilde{x}, \quad y = \underline{y} + (\overline{y} - \underline{y})\tilde{y}.$$

Then $f(x, y) = \tilde{X}(EAR^T)\tilde{Y}^T$ where

$$\tilde{X} = (1, \tilde{x}, ..., \tilde{x}^n), \quad \tilde{Y} = (1, \tilde{y}, ..., \tilde{y}^n),$$

$$E_{ij} = \begin{cases} \binom{j}{i}\underline{x}^{j-i}(\overline{x} - \underline{x})^i, & i \leq j \\ 0, & i > j \end{cases}, \quad i = 0, 1, ..., n; \quad j = 0, 1, ..., n,$$

and

$$R_{ij} = \begin{cases} \binom{j}{i}\underline{y}^{j-i}(\overline{y} - \underline{y})^i, & i \leq j \\ 0, & i > j \end{cases}, \quad i = 0, 1, ..., m; \quad j = 0, 1, ..., m.$$

Setting $G = EAR^T$, we complete the range conversion using

$$\tilde{f}(\tilde{x}, \tilde{y}) = \tilde{X}G\tilde{Y}^T, \quad (\tilde{x}, \tilde{y}) \in [0, 1] \times [0, 1].$$

Next we need to convert the above polynomial from the power basis to the Bernstein basis. (This is well-known to be an ill-conditioned problem, but as we will be computing conservative bounds using intervals, we can still arrange to guarantee the *correctness* of the results. Ill–effects will show up as fatter curves when plotting.)

Let

$$\begin{aligned} \tilde{B}(\tilde{X}) &= (B_0^n(\tilde{x}), B_1^n(\tilde{x}), ..., B_n^n(\tilde{x})), \\ \tilde{B}(\tilde{Y}) &= (B_0^m(\tilde{y}), B_1^m(\tilde{y}), ..., B_m^m(\tilde{y})), \end{aligned}$$

where $B_j^i(u) = \binom{i}{j}u^j(1-u)^{i-j}$ are the Bernstein basis functions. Then

$$\tilde{B}(\tilde{X}) = \tilde{X}H, \quad \tilde{B}(\tilde{Y}) = \tilde{Y}P,$$

where

$$H_{ij} = \begin{cases} 0, & i < j \\ (-1)^{i-j}\binom{n}{j}\binom{n-j}{i-j}, & i \geq j \end{cases}, \quad i = 0, 1, ..., n; \quad j = 0, 1, ..., n,$$

$$P_{ij} = \begin{cases} 0, & i < j \\ (-1)^{i-j}\binom{m}{j}\binom{m-j}{i-j}, & i \geq j \end{cases}, \quad i = 0, 1, ..., m; \quad j = 0, 1, ..., m.$$

Then
$$\tilde{f}(\tilde{x}, \tilde{y}) = \tilde{B}(\tilde{X})H^{-1}G(P^T)^{-1}\tilde{B}(\tilde{Y})^T,$$
so, letting $Q = H^{-1}G(P^T)^{-1}$, we obtain
$$\tilde{f}(\tilde{x}, \tilde{y}) = \tilde{B}(\tilde{X})Q\tilde{B}(\tilde{Y})^T, \quad (\tilde{x}, \tilde{y}) \in [0, 1] \times [0, 1].$$

The conversion from the power basis to Bernstein basis is completed.

By the Bernstein convex hull property [7] we know that
$$\underline{F} \leq \tilde{f}(\tilde{x}, \tilde{y}) \leq \overline{F}, \quad (\tilde{x}, \tilde{y}) \in [0, 1] \times [0, 1],$$
$$\underline{F} = \min_{i,j}\{Q_{ij}\}, \quad \overline{F} = \max_{i,j}\{Q_{ij}\}, \quad i \in \{0, 1, ..., n\}, \quad j \in \{0, 1, ..., m\},$$
and so
$$\underline{F} \leq f(x, y) \leq \overline{F}, \quad (x, y) \in [\underline{x}, \overline{x}] \times [\underline{y}, \overline{y}],$$
giving the desired bounds of $f(x, y)$ on range $[\underline{x}, \overline{x}] \times [\underline{y}, \overline{y}]$.

6. Examples

Here, we briefly compare the accuracy and speed of each method mentioned above using two examples. In each case we plot a curve $f(x, y) = 0$ using the algorithm described in Section 2, using a grid of 256×256 pixels. In the tables of results, we use the following notation: IAP, IAB (interval arithmetic using the power or Bernstein basis), AA (affine arithmetic matrix method), BC (Bernstein coefficient method).

The first example, from [18], shown in Figures 12.1–12.4, plots the curve $0.945xy - 9.43214x^2y^3 + 7.4554x^3y^2 + y^4 - x^3 = 0$ on $[0, 1] \times [0, 1]$.

The second example, from [19], shown in Figures 12.5–12.8, plots the curve $-1801/50 + 280x - 816x^2 + 1056x^3 - 512x^4 + 1601/25y - 512xy + 1536x^2y - 2048x^3y + 1024x^4y = 0$ on $[0, 1] \times [0, 1]$.

To compare the methods, various quantities were measured:

- The percentage of the overall area definitely classified as not containing the curve: the bigger the better.

- The number of subdivisions needed: the lower the better, because of overheads incurred in recursion.

- The number of additions (and subtractions), and multiplications needed: the lower the better.

Tables 12.1 and 12.2 give details of these quantities for both test cases.

Algebraic Methods for Curve Drawing

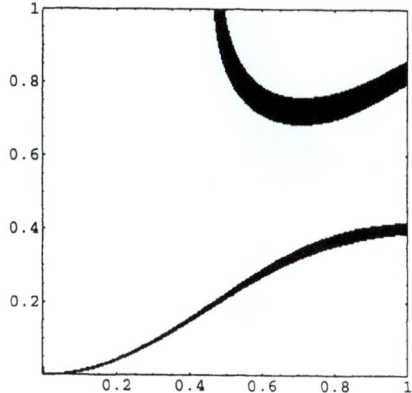

Figure 12.1. Example 1 drawn using IA and the power basis.

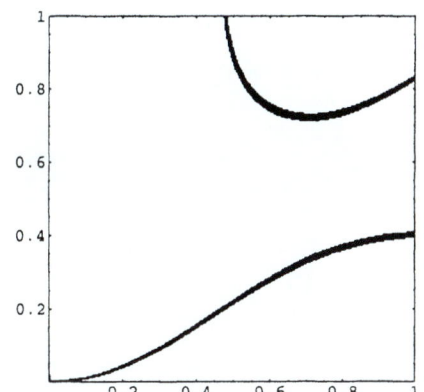

Figure 12.2. Example 1 drawn using IA and the Bernstein basis.

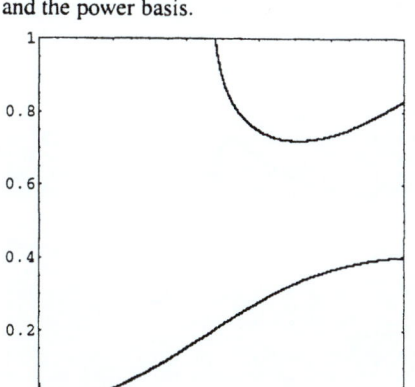

Figure 12.3. Example 1 drawn using AA.

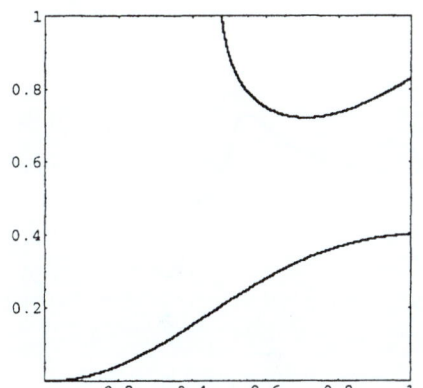

Figure 12.4. Example 1 drawn using BC.

Table 12.1. Comparison of IAP, IAB, AA and BC methods for the first example.

Methods	Area Classified	Subdivisions	Additions	Multiplications
IAP	93.8568	3909	1493798	1876438
IAB	97.5906	1564	2149336	2615424
AA	99.0723	634	1355919	870189
BC	99.0967	585	1429698	1108000

We consider the accuracy first. As can be seen from the percentage of area definitely classified, IAB is better than IAP, AA is better than IAB (sometimes much better as shown in Example 2), and BC is slightly better than AA.

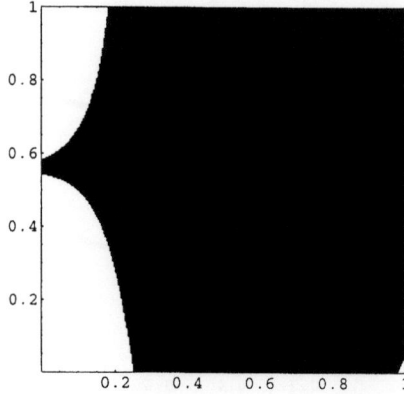

Figure 12.5. Example 2 drawn using IA and the power basis.

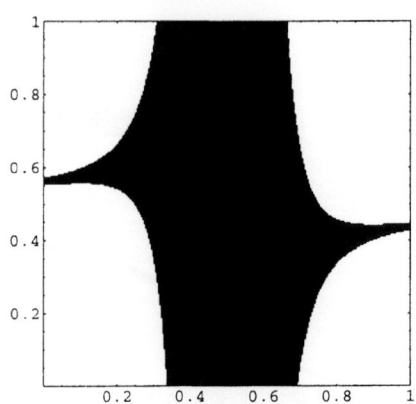

Figure 12.6. Example 2 drawn using IA and the Bernstein basis.

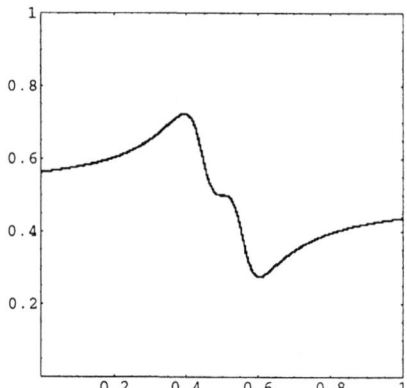

Figure 12.7. Example 2 drawn using AA.

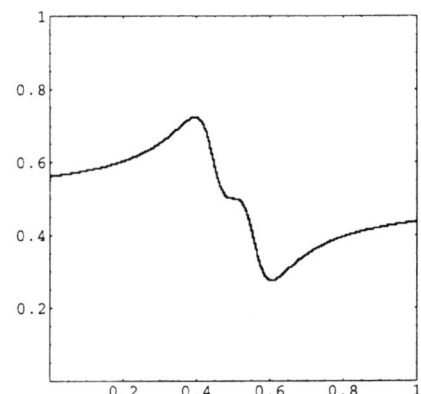

Figure 12.8. Example 2 drawn using BC.

Table 12.2. Comparison of IAP, IAB, AA and BC methods for the second example.

Methods	Area Classified	Subdivisions	Additions	Multiplications
IAP	15.8356	19562	9336638	11580850
IAB	53.9001	12735	11862746	14161596
AA	99.292	611	660781	430318
BC	99.3042	465	576289	409917

Next we consider the number of arithmetic operations. In general, there is no definite order in terms of the number of operations (additions and multiplications) involved. However, we notice IAB behaves worst, probably because of the complexity of the expressions involved in the examples. For simple curves like the first, the performance of the AA and BC methods is generally similar to that for the IAP method, although sometimes IAP can be several times faster. However, for more complicated curves, the AA and BC methods are often substantially faster—over 20 times faster in the case of the second curve.

7. Conclusions

Firstly, we have shown that the choice of polynomial basis does not affect results obtained with AA, unlike what happens with IA. Secondly, we conclude from the above examples, and other experiments we have carried out, that the AA and BC methods are roughly similar in accuracy, and generally more accurate than IAP and IAB methods. Furthermore, while there is sometimes a time penalty for using these methods, in tricky cases, they are generally faster than the IA methods.

Acknowledgments

We would like to thank the China Scholarship Council for funding Huahao Shou. We would also like to thank Bath University and the ORS award scheme for funding Irina Voiculescu.

References

[1] Berchtold, J., *The Bernstein Form in Set-Theoretic Geometric Modelling*, PhD Thesis, University of Bath, 2000.

[2] Berchtold, J., Bowyer, A., Robust Arithmetic for Multivariate Bernstein-Form Polynomials, *Computer Aided Design*, 2000, 32: 681–689.

[3] Bowyer, A., Berchtold, J., Eisenthal, D., Voiculescu, I., and Wise, K., Interval Methods in Geometric Modelling, *Geometric Modeling and Processing 2000*, Eds. Martin, R., Wang, W., IEEE Computer Society Press, 2000, 321–327.

[4] Chandler, R. E., A Tracking Algorithm for Implicitly Defined Curves, *IEEE Computer Graphics & Applications*, 1988, 8(2): 83–89.

[5] Comba, J. L. D., Stolfi, J., Affine Arithmetic and its Applications to Computer Graphics, *Anais do VII SIBGRAPI (Brazilian Symposium on Computer Graphics and Image Processing)*, Recife, Brazil, 1993, 9–18.

[6] Duff, T., Interval Arithmetic and Recursive Subdivision for Implicit Functions and Constructive Solid Geometry, *Computer Graphics (SIGGRAPH'92 Proceedings)*, 1992, 26(2): 131–138.

[7] R. T. Farouki, V. T. Rajan, On the Numerical Condition of Polynomials in Bernstein Form, *Computer Aided Geometric Design*, 191–216, 1987.

[8] de Figueiredo, L. H., Surface Intersection Using Affine Arithmetic, *Proceedings of Graphics Interface'96*, Eds. MacKenzie S., Stewart J., Morgan Kaufmann, 1996, 168–175.

[9] de Figueiredo, L. H., Stolfi, J., Adaptive Enumeration of Implicit Surfaces with Affine Arithmetic, *Computer Graphics Forum*, 1996, 15(5): 287–296.

[10] Heidrich, W., Slusallek, P., Seidel, H. P., Sampling of Procedural Shaders using Affine Arithmetic, *ACM Transaction on Graphics*, 1998, 17(3): 158–176.

[11] Moore, R. E., *Methods and Applications of Interval Analysis*, Society for Industrial and Applied Mathematics, Philadelphia, 1979.

[12] Mudur, S. P., Koparkar, P. A., Interval Methods for Processing Geometric Objects, *IEEE Computer Graphics & Applications*, 1984, 4(2): 7–17.

[13] Ratschek, H., Rokne, J., *Computer Methods for the Range of Functions*, Ellis Horwood Ltd., 1984.

[14] Shou H., Martin R., Voiculescu I., Bowyer A., Wang G., Affine Arithmetic in Matrix Form for Algebraic Curve Drawing, Progress in Natural Science, 12 (1) 77–81, 2002.

[15] Snyder, J. M., Interval Analysis for Computer Graphics, *Computer Graphics (SIGGRAPH'92 Proceedings)*, 1992, 26(2): 121–130.

[16] Suffern, K. G., Quadtree Algorithms for Contouring Functions of Two Variables, *The Computer Journal*, 1990, 33: 402–407.

[17] Suffern, K. G., Fackerell, E. D., Interval Methods in Computer Graphics, *Computer & Graphics*, 1991, 15: 331–340.

[18] Voiculescu, I., Berchtold, J. Bowyer, A. Martin, R. R., Zhang, Q., Interval and Affine Arithmetic for Surface Location of Power- and Bernstein-form Polynomials, *The Mathematics of Surfaces IX*, Eds. Cipolla R. & Martin, R. R., Springer, 2000, 410–423.

[19] Zhang, Q., Martin, R. R., Polynomial Evaluation using Affine Arithmetic for Curve Drawing, *Proceedings of Eurographics UK 2000*, Eurographics UK, Abingdon, 2000, 49–56.

Chapter 13

LOCAL POLYNOMIAL METRICS FOR K NEAREST NEIGHBOR CLASSIFIERS

Robert R. Snapp
Department of Computer Science
University of Vermont
Burlington, VT 05405, USA
snapp@cs.uvm.edu

Keywords: k-nearest-neighbor classifier, local polynomial model, metric, pattern recognition

1. Introduction

In many pattern recognition systems, metrics are frequently employed to quantify the dissimilarity that exists between two given patterns. Common applications occur in clustering algorithms, radial basis function classifiers, and nearest neighbor classifiers (Duda et al., 2001). A bounty of results exist that demonstrate the importance of selecting a metric (or dissimilarity measure) with care (e.g., Simard et al., 1993). In the following, we present an adaptive algorithm that uses local polynomial regression to construct a metric useful for a pattern classification problem described by a set of correctly classified patterns. Although we introduce this algorithm in the context of the k nearest neighbor classifier, it is applicable to other pattern classifiers that use proximity in feature space as a measure of pattern similarity.

We assume that each pattern originates from a unique state of nature, indicated by a discrete *class label*, $\ell \in \{1, \ldots, C\}$, and that each pattern is represented by a finite-dimensional *feature vector*, $\mathbf{x} = (x_1, \ldots, x_d) \in \mathbf{R}^d$. Feature selection is an art that benefits from a deep understanding of the underlying structure of each pattern recognition problem. If the features are well chosen, and the underlying pattern recognition problem possesses some degree of regularity, then the distance between two feature vectors can serve as a measure of pattern dissimilarity.

A popular choice is a weighted L_p metric, $D(\mathbf{x}, \mathbf{y}) = \|W(\mathbf{x} - \mathbf{y})\|_p$ where

$$\|\mathbf{x}\|_p \equiv \begin{cases} \sqrt[p]{(|x_1|^p + \cdots + |x_d|^p)} & : \text{ if } 1 \leq p < \infty, \\ \max_{1 \leq i \leq d} |x_i| & : \text{ if } p = \infty, \end{cases}$$

denotes the L_p norm, and $W \in \mathbf{R}^{d \times d}$ represents a weight matrix. If $p = 2$ and $W = I$ (the identity matrix), then the above describes the Euclidean metric. Similarly, $p = 1$ corresponds to the Manhattan metric, and $p = \infty$ to the supremum metric.

In this study, our goal is to discover a weighted L_p metric that enhances the accuracy of k nearest neighbor classifiers. Asymptotic analysis suggests that a weighted Euclidean metric ($p = 2$) is an optimal L_p metric if the probability distributions that describe the classification problem are sufficiently smooth (Snapp and Venkatesh, 1998). Thus, in the following we consider metrics of the form

$$D(\mathbf{x}, \mathbf{y}) = \sqrt{(\mathbf{x} - \mathbf{y})^T A (\mathbf{x} - \mathbf{y})}, \qquad (13.1)$$

where A is a $d \times d$ matrix that can vary slowly over the feature space, and $\mathbf{x}^T$ denotes the transpose of $\mathbf{x}$.

2. k nearest neighbor classifiers

After fifty years, the k-nearest-neighbor classifier continues to serve as an important pattern recognition paradigm (Fix and J. L. Hodges, 1951). Using this algorithm, an input pattern (represented by the feature vector $\mathbf{x} \in \mathbf{R}^d$) is assigned to a pattern class with the aid of a reference sample of n previously classified patterns. In its classic form, each input vector is assigned to the class represented by the majority of the k reference vectors (for a given integer k) that are closest to the input vector. (Ties can be broken by an arbitrary procedure.)

The reference sample is usually represented as a set $\mathcal{S}_n$ of n ordered pairs,

$$\mathcal{S}_n = \left\{ \left(\mathbf{x}^{(1)}, \ell^{(1)}\right), \left(\mathbf{x}^{(2)}, \ell^{(2)}\right), \ldots, \left(\mathbf{x}^{(n)}, \ell^{(n)}\right) \right\}, \qquad (13.2)$$

where the first component, $\mathbf{x}^{(j)} \in \mathbf{R}^d$ is a feature vector, and the second, $\ell^{(j)} \in \{1, 2, \ldots, C\}$ is the correct class label of $\mathbf{x}^{(j)}$.

Given an input feature $\mathbf{x}$, the k nearest neighbor classifier identifies the subset of k nearest neighbors of $\mathbf{x}$: $\left\{ \left(\mathbf{x}^{(i_1)}, \ell^{(i_1)}\right), \left(\mathbf{x}^{(i_2)}, \ell^{(i_2)}\right), \ldots, \left(\mathbf{x}^{(i_k)}, \ell^{(i_k)}\right) \right\}$ where $i_1, i_2, \ldots, i_k$ are chosen without replacement from the set $\{1, 2, \ldots, n\}$, such that

$$D\left(\mathbf{x}, \mathbf{x}^{(i_1)}\right) \leq D\left(\mathbf{x}, \mathbf{x}^{(i_2)}\right) \leq \cdots \leq D\left(\mathbf{x}, \mathbf{x}^{(i_k)}\right) \leq D\left(\mathbf{x}, \mathbf{x}^{(j)}\right),$$

Local polynomial metrics for k nearest neighbor classifiers 157

for all $j \in \{1, 2, \ldots, n\} \setminus \{i_1, i_2, \ldots, i_k\}$. Finally, the input vector **x** is assigned to the class that appears most frequently in the subset $\{\ell^{(i_1)}, \ell^{(i_2)}, \ldots, \ell^{(i_k)}\}$.

In effect, the k nearest neighbor algorithm empirically estimates the posterior probability $\mathbf{P}\{\ell|\mathbf{x}\}$ of class ℓ (that is, the probability that the given feature vector **x** originates from class ℓ) as the relative frequency of label ℓ within the subset of k nearest neighbors. It then assigns the input pattern **x** to the class with the greatest estimated posterior probability. Under weak assumptions, this algorithm achieves a Bayes classifier in the infinite-sample limit: $n \to \infty, k \to \infty$ such that $k/n \to 0$. In practice, however, the reference sample and k are finite. Thus the accuracy of this classifier depends on the choice of k and on the metric. With the reference sample, these elements determine the shape and size of the neighborhood of **x**, $\mathcal{N}(\mathbf{x})$, that is analyzed. For D given by Eqn. (13.1), the neighborhood is ellipsoidal, with principal axes described by the eigenvectors and eigenvalues of the weight matrix A. The shape of this neighborhood can bias the estimates of the posterior probabilities if the true values vary significantly over the extent of $\mathcal{N}(\mathbf{x})$. As k increases, the variance of each estimate decreases. However the size of the neighborhood increases, as does the resulting bias.

3. Minimum bias heuristic

We now describe the method for selecting a weight matrix A in Eqn. (13.1) that enhances the accuracy of a k nearest neighbor classifier. Let

$$\mathbf{P}(\mathbf{y}) = (\mathbf{P}\{1|\mathbf{y}\}, \mathbf{P}\{2|\mathbf{y}\}, \ldots, \mathbf{P}\{C|\mathbf{y}\})^T \in [0, 1]^C, \quad (13.3)$$

denote a C dimensional vector that contains the posterior probabilities of each of the C classes as components, and assume that A in Eqn. (13.1) represents a positive definite, $d \times d$ matrix, with $\det A = 1$. (If $\det A \neq 1$, then dividing each element of A by $\sqrt[d]{\det A}$ yields a normalized weight matrix that induces an equivalent metric.) Given a reference sample, $\mathcal{S}_n$, and an input pattern **x**, a k nearest neighbor classifier will analyze the feature vectors that lie within the elliptical neighborhood

$$\mathcal{N}(\mathbf{x}) = \{\mathbf{y} \in \mathbf{R}^d : D(\mathbf{x}, \mathbf{y}) \leq \rho_k\},$$

where $\rho_k \equiv D\left(\mathbf{x}, \mathbf{x}^{(i_k)}\right)$ denotes the distance between **x** and its k-th nearest neighbor. This ellipsoid is centered at **x**, and its shape depends upon the choice of A. Let

$$E_A = \{\mathbf{y} \in \mathbf{R}^d : D(\mathbf{x}, \mathbf{y}) \leq 1\},$$

denote a similar ellipsoid of standard volume, $V = \pi^d / \Gamma\left(\frac{d+2}{2}\right)$, centered about **x**.

By varying A (subject to the constraint that $\det A = 1$), we seek an ellipsoid within which the variation of the posterior probabilities,

$$\psi[A] = \frac{1}{V}\int_{E_A} \|\mathbf{P}(\mathbf{y}) - \mathbf{P}(\mathbf{x})\|_2^2 \, d\mathbf{y}, \qquad (13.4)$$

is minimized. Using this heuristic, one obtains a weight matrix that yields a similarity measure that favors nearest neighbors that have a posterior probability distribution similar to that at the observation point $\mathbf{x}$. This reduces the bias error of the posterior probability estimates. Similar heuristics are advocated by Friedman, 1994, and by Hastie and Tibshirani, 1996. (The DANN algorithm, derived in the latter work, uses an iterative discriminant analysis to identify the directions in feature space along which the neighborhood should be elongated.)

4. Local polynomial models of $P\{\ell|\mathbf{x}\}$

In practice, the posterior probabilities are not known. However, they can be estimated directly from the reference sample S_n defined in Eqn. (13.2) using *local polynomial models* (see Fan and Gijbels, 1996). Given $\mathbf{x}$, we construct a local quadratic model that approximates $P\{\ell|\mathbf{y}\}$, the (posterior) probability that pattern $\mathbf{y}$ belongs to class ℓ, for patterns $\mathbf{y}$ in the neighborhood of $\mathbf{x}$. For $\ell = 1, 2, \ldots, C$, let $g_\ell \in \mathbf{R}$, $\mathbf{g}_\ell \in \mathbf{R}^d$, and $G_\ell \in \mathbf{R}^{d\times d}$ denote the coefficients that minimize the sum of weighted residuals,

$$\sum_{i=1}^n \left(\left(\delta_{\ell,\ell^{(i)}} - \left(g_\ell + \mathbf{g}_\ell^T(\mathbf{x}^{(i)} - \mathbf{x}) + (\mathbf{x}^{(i)} - \mathbf{x})^T G_\ell(\mathbf{x}^{(i)} - \mathbf{x})\right)\right)\right)^2$$
$$\times K_h(\|\mathbf{x}^{(i)} - \mathbf{x}\|_2)), \qquad (13.5)$$

where $\delta_{\ell,\ell'}$ denotes a Kronecker delta function, and $K_h(\xi)$ is a kernel with bandwidth $h > 0$, e.g.,

$$K_h(\xi) = \frac{1}{(2\pi)^{d/2}\,h}\exp\left(-\frac{\xi^2}{2h^2}\right).$$

The coefficients g_ℓ, $\mathbf{g}_\ell$, and G_ℓ can be found algebraically, using the method of least squares. To avoid numerical instability, the size of the reference sample should satisfy $n > (d+2)(d+1)/2$. Ill-conditioning can also occur if the bandwidth h is too small. Accordingly, h is often determined by an adaptive algorithm. Further technical details can be found in (Fan and Gijbels, 1996).

The local quadratic model of $P\{\ell|\mathbf{y}\}$ is then

$$\mathbf{P}\{\ell|\mathbf{y}\} \approx g_\ell + \mathbf{g}_\ell^T(\mathbf{y} - \mathbf{x}) + (\mathbf{y} - \mathbf{x})^T G_\ell(\mathbf{y} - \mathbf{x}). \qquad (13.6)$$

Substituting Eqn. (13.6) into the objective function, Eqn. (13.4), yields

$$\psi[A] = \sum_{\ell=1}^{C} \left(\frac{\mathbf{g}_\ell^T A^{-1} \mathbf{g}_\ell}{d+2} + 2\frac{\text{tr}((A^{-1}G_\ell)^2)}{(d+2)(d+4)} + \frac{(\text{tr } A^{-1}G_\ell)^2}{(d+2)(d+4)} \right). \quad (13.7)$$

Let $A^\star$ denote a value of A that minimizes Eqn. (13.7) subject to the normalization constraint $\det A = 1$. The value of $A^\star$ is readily obtained using a quasi-Newton optimization procedure (see Bertsekas, 1995). To enforce the positivity and normalization of A, we represent A^{-1} as the product $LL^T(\det L)^{-2/d}$, where L denotes an unspecified lower-triangular matrix. Then, the optimal weight matrix $A^\star$ can be obtained by minimizing ψ with respect to L in an unconstrained fashion.

The above expression for ψ illustrates the need for expanding the posterior probabilites out to second order: if $C < d$ and the G_ℓ term in Eqn. (13.4) is neglected, then the resulting elliptical neighborhood is undefined.

In summary, each pattern classification requires three steps. First, a local polynomial model for each posterior probability $\mathbf{P}\{\ell|\mathbf{y}\}$ is constructed about the input vector $\mathbf{x}$. Then, the weight matrix $A^\star$ that minimizes the objective function $\psi[A]$ is computed. Finally, $\mathbf{x}$ is assigned to a class by a k nearest neighbor classifier that assumes the weighted Euclidean metric defined by $A^\star$. The same reference sample, $\mathcal{S}_n$, can be used for the first and last steps. The entire procedure has been implemented in *Mathematica*. In practice, determining the coefficients of the local polynomial approximation dominates the execution, with time complexity of order $O\left(d^4(n+d^2)\right)$. The time complexity of the optimization step is more difficult to analyze, but the value $L = I$ appears to be a robust initial condition. The utility of this algorithm is demonstrated by the following examples.

5. Example: Three normally distributed classes

Consider an artificial three class problem described by feature vectors generated from three normal distributions, each centered on a vertex of an equilateral triangle, each with unit variance. Let $d = 2$, $C = 3$, and assume that the class-conditional densities of the three classes are

$$p(\mathbf{x}|\ell) = \frac{1}{2\pi} e^{-\|\mathbf{x}-\mu_\ell\|^2/2},$$

for $\ell = 1, 2, 3$, where

$$\mu_1 = (1, 0), \quad \mu_2 = \left(-\frac{1}{2}, \frac{\sqrt{3}}{2}\right), \quad \mu_3 = \left(-\frac{1}{2}, -\frac{\sqrt{3}}{2}\right),$$

are the vertices of an equilateral triangle. Assume the prior probabilities of each class are equal: $P_1 = P_2 = P_3 = 1/3$. Using a pseudo-random number

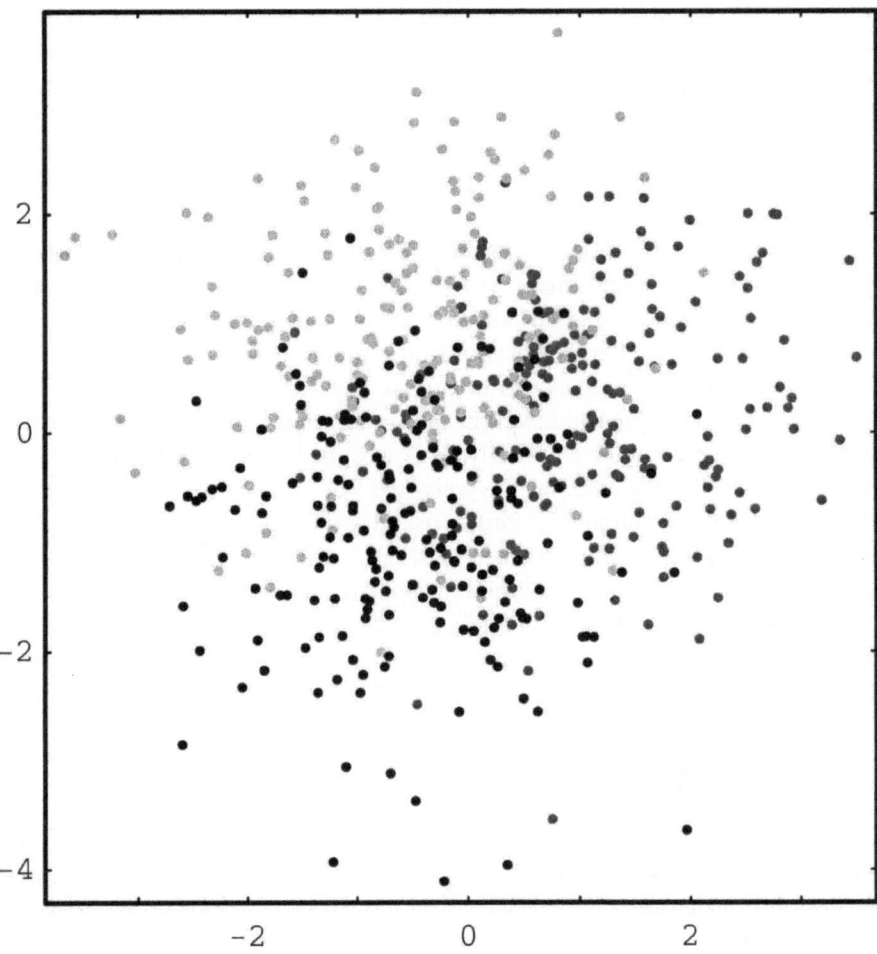

Figure 13.1. Six-hundred labeled patterns in $\mathbf{R}^2$, chosen at random for a three-class problem, in which each class is generated from a normal distribution with a mean coincident on a vertex of an equilateral triangle.

Local polynomial metrics for k nearest neighbor classifiers

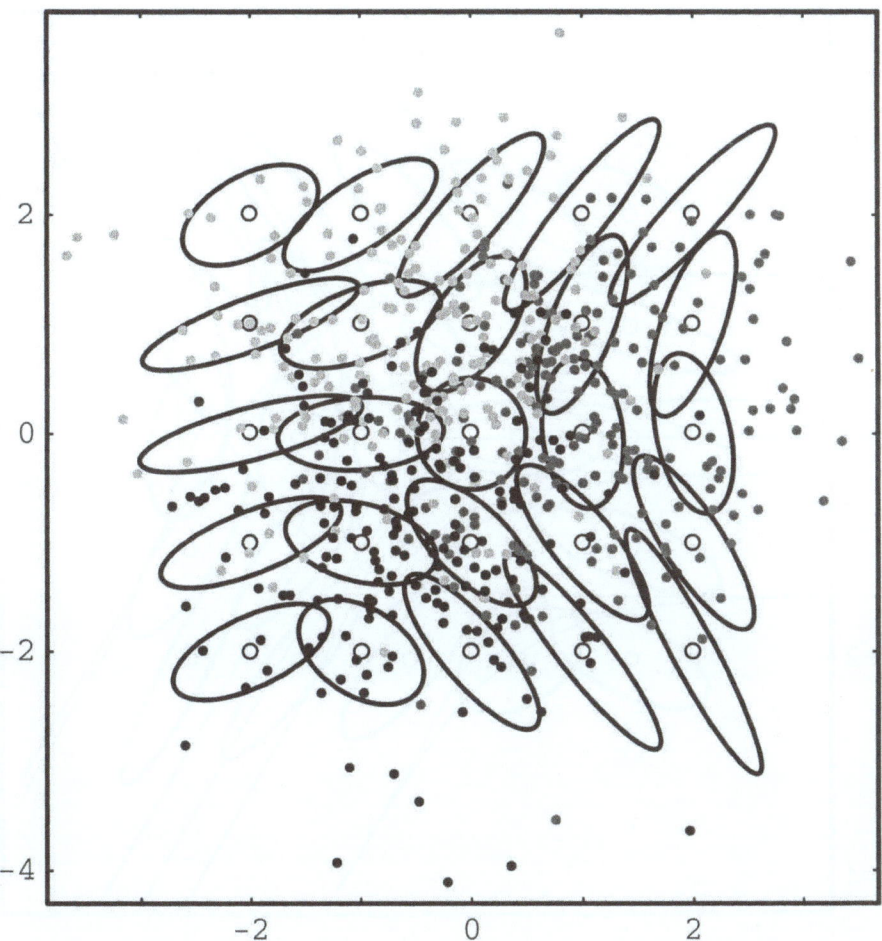

Figure 13.2. Twenty-five elliptical neighborhoods obtained by minimizing the objective function $\psi[A]$ at input feature vectors that lie in a 5×5 lattice.

generator, 600 labeled feature vectors were generated at random, and appear in Figure 13.1 in the color that displays their class label.

Next, twenty-five input patterns **x** in a 5×5 square grid are analyzed. For each point, the preceding three-step algorithm is applied using only the information contained in the reference sample of 600 points. The resulting elliptical neighborhoods are displayed as gray ovals in Figure 13.2.

Since this is an artificial problem, with a known probability distribution, it is possible to compute the posterior probabilities exactly. Figure 13.3 illus-

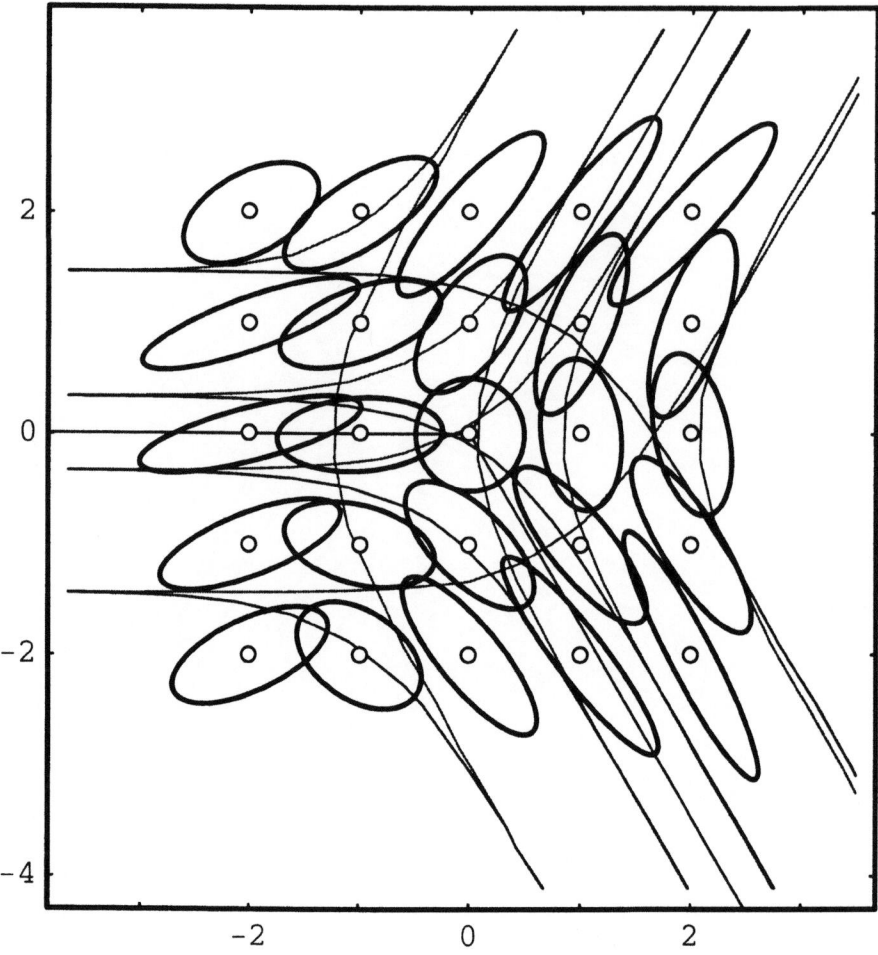

Figure 13.3. The same twenty-five elliptical neighborhoods superimposed on the contours of constant posterior probability $\mathbf{P}\{\ell|\mathbf{y}\}$ for each of the three pattern classes.

trates the contours of constant posterior probability for each class with the 25 elliptical neighborhoods.

With the exception of the neighborhoods centered at $(-2, 0)$ and $(-2, -1)$, the ellipses are elongated along the iso-contours of the posterior probability of the most probable class. Note also that the neighborhood centered at $(0, 0)$ is circular, as symmetry demands.

6. Example: The Pima Indian Diabetes Problem

In this experiment we apply this local metric to the Pima Indian data set. The original data set is available from the UC Irvine Machine Learning Repository created by (Blake and Merz, 1998). (Ripley, 1996) has used this data set as a test bed to illustrate the differences between a variety of pattern classifiers. In the following, we assume the same partition as Ripley, i.e., a reference sample of $n = 200$ patterns, and 332 test patterns. Each raw seven-dimensional feature vector is projected into a four-dimensional feature space by dropping the first, third, and fourth attributes. Each remaining component is normalized by dividing it by its variance within the reference sample. For odd values of k, 1 through 23, we construct two k nearest neighbor classifiers: one with $A = I$ (an identity matrix), the other with $A = A^*$ obtained from our procedure. The number of misclassifications obtained in each experiment are shown in Table 13.1. For each value of k, the smallest number of misclassifications is shown in boldface. Note that the local method performs significantly better on this data set. Moreover, for $k = 13$ the accuracy equals that of the best algorithm described in (Ripley, 1996).

7. Conclusion

Our experiments demonstrate that the use of a local metric derived by the above algorithm can improve the accuracy of a k nearest neighbor classifier. Although these results are preliminary, they demonstrate that local polynomial models can be used to design useful metrics. Of perhaps greater significance is the suggestion that flexible metric design can be used to complement the difficult task of feature selection, as a weight matrix A with some negligible eigenvalues can discount the influence of improperly chosen features. Thus, algorithms that discover effective metrics from sample data may benefit a wide variety of pattern recognition systems.

Table 13.1. Number of misclassifications obtained for twelve values of k using a uniform Euclidean metric ($A = I$), and a local metric obtained by the three step procedure ($A = A^*$).

k	1	3	5	7	9	11	13	15	17	19	21	23
$A = I$	100	90	79	79	78	74	77	74	76	**72**	**68**	74
$A = A^*$	**85**	**82**	**76**	**72**	**71**	**69**	**64**	**68**	**71**	74	71	**69**

Acknowledgments

The author would like to than Dr. Ron Meir. This research was funded by the U.S. Army Research Office under grant USARO DAAG55-9810022.

References

Bertsekas, D. (1995). *Nonlinear Programming*. Athena Scientific, Belmont, MA.

Blake, C. and Merz, C. (1998). UCI repository of machine learning databases.

Duda, R. O., Hart, P. E., and Stork, D. G. (2001). *Pattern Classification*. Wiley, New York, second edition.

Fan, J. and Gijbels, I. (1996). *Local Polynomial Modelling and Its Applications*. Chapman & Hill, London.

Fix, E. and J. L. Hodges, J. (1951). Discriminatory analysis — nonparametric discrimination: consistency properties. Technical Report 4, USAF School of Aviation Medicine, Randolf Field, TX. Project 21-49-004.

Friedman, J. (1994). Flexible metric nearest neighbor classification. Technical report, Stanford University, Department of Statistics.

Hastie, T. and Tibshirani, R. (1996). Discriminant adaptive nearest neighbor classification. *IEEE Transactions on Pattern Analysis and Machine Intelligence*, 18:607–615.

Ripley, B. D. (1996). *Pattern Recognition and Neural Networks*. Cambridge University Press, Cambridge.

Simard, P., Cun, Y. L., and Denker, J. (1993). Efficient pattern recognition using a new transformation distance. In Giles, C. L., editor, *Advances in Neural Information Processing Systems*, pages 50–58, San Mateo, CA. Morgan Kaufmann.

Snapp, R. R. and Venkatesh, S. S. (1998). Asymptotic expansions of the k nearest neighbor risk. *Annals of Statistics*, 26:850–878.

Chapter 14

VISUALISATION OF INCOMPLETE DATA USING CLASS INFORMATION CONSTRAINTS

Yi Sun, Peter Tino, Ian Nabney
Neural Computing Research Group
Aston University, Birmingham B4 7ET, United Kingdom
{suny,tinop,i.t.nabney}@aston.ac.uk

Abstract We analyse how the training algorithm for the Generative Topographic Mapping (GTM) can be modified to use class information to improve results on incomplete data. The approach is based on an Expectation-Maximisation (EM) method which estimates the parameters of the mixture components and missing values at the same time; furthermore, if we know the class membership of each pattern, we can improve the generic algorithm by eliminating multi-modalities in the posterior distribution over the latent space centres. We evaluate the method on a toy problem and a realistic data set. The results show that our algorithm can help to construct informative visualisation plots, even when many of the training points are incomplete.

Keywords: Incomplete data, GTM, EM, class information

1. Introduction

Data visualisation plays a key role in developing good models for large quantities of data and can aid analysts in identifying groups of data points which have similar characteristics. It is often the case that the visualisation plot can be more helpful when data is labelled with class information (using colour, for example), and in many datasets, this class information is readily available.

In many applications the input data is incomplete. Therefore it is important to use available values and class information and to reconstruct the missing values. For example, in the field of drug discovery, scientists use computer modelling to analyse the molecular structure of compounds and high throughput screening to assess their interaction with biological targets. Many compounds are not screened against a complete set of targets, yet we do not want to exclude all such compounds from data analysis since that risks missing potential drugs.

Tresp et al. ([9] and [8]) developed techniques for coping with missing data in the context of supervised learning. In this paper, we investigate ways of dealing with missing data in the context of unsupervised and "semi-supervised" learning. In particular, we concentrate on the generative topographic mapping GTM [2], which is a non-linear probabilistic visualisation technique. The problem considered here is how to train the GTM model with incomplete data and reconstruct the missing values using additional class information. In this way the data, including the missing components, can be shown in a visualisation plot that is as "faithful" as possible.

Our algorithm can be described briefly as follows. A density model (GTM) of the data is learned in an unsupervised way from the incomplete training data set using an EM algorithm where sufficient statistics for the missing data are estimated in the E-step. For visualisation purposes, the missing values are filled in by computing the mean of their conditional distribution. The generic algorithm [6] can be improved by considering class label information associated with each point in the training data set. In this case, we can obtain the probability that a particular class c was responsible for generating the latent space centre. By using Bayes' theorem, we can compute the class-conditional priors for all latent space centres, which can be imposed on the latent space centres when we calculate the posterior probabilities (or 'responsibilities') of the latent space centres for the incomplete data points.

The generic algorithm, GTM with incomplete data, is detailed in section 2. Section 3 gives a detailed description of how we incorporate class information into the generic algorithm. We illustrate the algorithm in section 4 with a toy data set and a high-dimensional data set. Section 5 discusses the results.

2. Generative Topographic Mapping

2.1 The Generative Topographic Mapping

The generative topographic mapping (GTM) [1] is a non-linear model that uses latent (or hidden) variables to model a probability distribution in the data space. It is a constrained mixture of Gaussians whose parameters are optimised using the expectation-maximisation (EM) algorithm [4].

For the GTM, $\mathbf{t}$ denotes the data in a D-dimensional Euclidean space and $\mathbf{x}$ denotes the latent variables in an L-dimensional latent space. A latent data point $\mathbf{x}$ is mapped to $\mathbf{y} = \mathbf{W}\boldsymbol{\Phi}(\mathbf{x})$ in data space by a radial basis function network (see e.g. [3]) with weights $\mathbf{W}$ and a basis function matrix $\boldsymbol{\Phi}$. The goal of the model is to represent the distribution $P(\mathbf{t})$ in terms of K latent points $\mathbf{x}_j$ ($j = 1, 2, ..., K$) and corresponding spherical Gaussian kernels centred on

$y(x_j; W)$ [2]. The data density is defined by

$$P(t|W, \beta) = \frac{1}{K} \sum_{j=1}^{K} P(t|x_j, W, \beta), \quad (14.1)$$

where $P(t|x_j, W, \beta)$ is the density given by the jth component, i.e.

$$P(t|x_j, W, \beta) = \left(\frac{\beta}{2\pi}\right)^{\frac{D}{2}} \exp\left\{-\frac{\beta}{2}\|t - y(x_j, W)\|^2\right\}.$$

The weights W and the inverse variance β can be fitted by maximum likelihood with the EM algorithm, while the mixing coefficients are fixed to $1/K$. The M-step includes solving linear equations for W and a simple weighted variance formula for β^{-1}.

The latent space representation of the point t_n, i.e. the *projection* of t_n, is taken to be the mean $\sum_{j=1}^{K} r_{nj} x_j$ of the posterior distribution on the latent space, where the responsibility r_{nj} is given by

$$r_{nj} = \frac{\left(\frac{\beta}{2\pi}\right)^{D/2} \exp\left\{-\frac{\beta}{2}\|y(x_j; W) - t_n\|^2\right\}}{\sum_{j'=1}^{K} \left(\frac{\beta}{2\pi}\right)^{D/2} \exp\left\{-\frac{\beta}{2}\|y(x_{j'}; W) - t_n\|^2\right\}}. \quad (14.2)$$

2.2 Incorporating missing values into the EM algorithm for the GTM model

To handle missing values in the data set, we write data points t_n as (t_n^o, t_n^m), where m and o denote subvectors and submatrices of the parameters matching the missing and observed components of the data, and each data vector can have different patterns of missing components. Binary indicator variables z_{nj} are introduced in the usual way to specify which component of the mixture generated the data point. $z_{nj} = 1$ if and only if t_n is generated by component j, otherwise $z_{nj} = 0$. The EM algorithm treats both the indicator variables z_{nj} and the missing inputs t_n^m as hidden variables.

For the GTM, in the E-step, the expectation of z_{nj} is $E[z_{nj}|t_n^o, \theta_j] = r_{nj}$ (equation (14.2)) measured only on t_n^o (the observed dimensions of t_n).

Following [5], we introduce

$$\hat{t}_{nj}^m = E(t_n^m|z_{nj} = 1, t_n^o, \theta_j) = (y_j^m)^{old} + \Sigma_j^{mo}\Sigma_j^{oo^{-1}}(t_n^o - (y_j^o)^{old}), \quad (14.3)$$

which is the least-squares regression between t_n^m and t_n^o predicted by Gaussian j, and 'old' denotes the value computed in the last M-step, $(y_j^m)^{old} = (W_{old}\Phi(x_j))^m$ and $(y_j^o)^{old} = (W_{old}\Phi(x_j))^o$. As the covariance matrix is isotropic, $\Sigma_j = \beta^{-1}\mathbf{I}$, and the covariance of missing and observed values Σ_j^{mo} is equal to 0. So we have:

$$\hat{\mathbf{t}}_{nj}^m = (\mathbf{y}_j^m)^{old}. \qquad (14.4)$$

The missing values are filled in using the posterior means,

$$E[\mathbf{t}_n^m|\mathbf{t}_n^o, \theta_j] = \sum_{j=1}^{K} r_{nj}\hat{\mathbf{t}}_{nj}^m. \qquad (14.5)$$

In the M-step, the new weights are updated to $\mathbf{W}_{new}$ as described in [2] for complete training data. The variance is updated by

$$\beta^{-1} = \frac{1}{ND}\sum_{n=1}^{N}\sum_{j=1}^{K} r_{nj}\left(\|\mathbf{t}_n^o - \mathbf{y}_j^o\|^2 + E[z_{nj}\|\mathbf{t}_n^m - \mathbf{y}_j^m\|^2]\right), \qquad (14.6)$$

where

$$\begin{aligned}E[z_{nj}\|\mathbf{t}_n^m - \mathbf{y}_j^m\|^2] &= M(\beta^{-1})^{old} + (\hat{\mathbf{t}}_{nj}^m)^T(\hat{\mathbf{t}}_{nj}^m) - 2(\hat{\mathbf{t}}_{nj}^m)^T\mathbf{y}_j^m \\ &\quad + (\mathbf{y}_j^m)^T\mathbf{y}_j^m,\end{aligned} \qquad (14.7)$$

where M is the number of missing values in $\mathbf{t}_n$, and $\mathbf{y}_j^o$ and $\mathbf{y}_j^m$ are computed using $(\mathbf{W}_{new}\Phi(\mathbf{x}_j))^o$ and $(\mathbf{W}_{new}\Phi(\mathbf{x}_j))^m$. A more detailed derivation can be found in [7].

3. Learning with Class-Conditional Information

When visualising a set of data points $T = \{\mathbf{t}_1, ..., \mathbf{t}_n\}$ from a set of class-labelled points $T_c = \{(\mathbf{t}_1, c_1), (\mathbf{t}_2, c_2), ..., (\mathbf{t}_N, c_N)\}$, with class labels c_n from a set $\mathcal{C} = \{\mathcal{C}_1, ..., \mathcal{C}_I\}$, one can use the class information as a clue for reasoning about missing values in the corrupted data points $\mathbf{t}_n$. Given a corrupted point $\mathbf{t}_n$, instead of computing the responsibilities $r_{nj} = P(\mathbf{x}_j|\mathbf{t}_n^o)$, we determine the *class-conditional responsibilities*

$$r_{njc} = P(\mathbf{x}_j|\mathbf{t}_n^o, c_n) = \frac{P(\mathbf{x}_j, c_n|\mathbf{t}_n^o)}{\sum_{k=1}^{K} P(\mathbf{x}_k, c_n|\mathbf{t}_n^o)}. \qquad (14.8)$$

By the (standard) assumption of conditional independence of observed variables, given the hidden ones, we have $P(\mathbf{t}_n^o, c_n|\mathbf{x}_j) = P(\mathbf{t}_n^o|\mathbf{x}_j)P(c_n|\mathbf{x}_j)$. Using a flat prior on the latent space centres $P(\mathbf{x}_k) = 1/K, k = 1, 2, ..., K$,

$$P(\mathbf{x}_j, c_n|\mathbf{t}_n^o) = \frac{P(\mathbf{t}_n^o, c_n|\mathbf{x}_j)P(\mathbf{x}_j)}{\sum_{k=1}^{K}\sum_{\mathcal{C}_i \in \mathcal{C}} P(\mathbf{t}_n^o, \mathcal{C}_i|\mathbf{x}_k)P(\mathbf{x}_k)} = r_{nj}P(c_n|\mathbf{x}_j). \qquad (14.9)$$

The distribution of class labels, conditioned on the latent space centres $\mathbf{x}_j$, is computed by determining the "mass" of uncorrupted training points "explained" by $\mathbf{x}_j$ and belonging to a class $C_i \in \mathcal{C}$,

$$P(C_i|\mathbf{x}_j) = \frac{\sum_{\mathbf{t}_n \in T_{comp}; c_n = C_i} P(\mathbf{t}_n|\mathbf{x}_j)}{\sum_{\mathbf{t}_n \in T_{comp}} P(\mathbf{t}_n|\mathbf{x}_j)}, \qquad (14.10)$$

where T_{comp} is a collection of uncorrupted training points. This means we are assuming that the mechanism for missing data does not depend on the class.

After some calculation, (14.8) can be rewritten as

$$r_{njc} = \frac{P(\mathbf{t}_n^o|\mathbf{x}_j) P(c_n|\mathbf{x}_j)}{\sum_{k=1}^{K} P(\mathbf{t}_n^o|\mathbf{x}_k) P(c_n|\mathbf{x}_k)} = \frac{P(\mathbf{t}_n^o|\mathbf{x}_j) P(\mathbf{x}_j|c_n)}{\sum_{k=1}^{K} P(\mathbf{t}_n^o|\mathbf{x}_k) P(\mathbf{x}_k|c_n)}, \qquad (14.11)$$

where for $C_i \in \mathcal{C}$,

$$P(\mathbf{x}_j|C_i) = \frac{P(C_i|\mathbf{x}_j)}{\sum_{k=1}^{K} P(C_i|\mathbf{x}_k)}. \qquad (14.12)$$

It follows from (14.11), that unlike the original latent centres' responsibilities r_{nj}, where a flat prior $P(\mathbf{x}_k) = 1/K$ is imposed, the class-conditional responsibilities, r_{njc}, are calculated using a *class-conditional prior on latent space*, $P(\mathbf{x}_j|C_i)$, $C_i \in \mathcal{C}$. When faced with corrupted data points, such class-conditional priors help us to eliminate (at least to some degree) multi-modalities in posterior over the latent space centres. This may be no longer a true EM algorithm. However, in our experiments there was no occasion on which the likelihood decreased, even with large numbers of missing values.

4. Experimental Results

In this section we compare three training algorithms for the GTM with missing data: **(I)** generic EM for missing data using class information as described in Section 14.3; **(II)** standard EM on the data first completed by class-conditional mean imputation. In this case, each missing component of a data point $\mathbf{t}_n$ from class C_i is filled by simply calculating the average of all observed values of the same component (dimension) in the points from the class C_i; **(III)** generic EM for missing data (i.e. with on-line estimation of missing values but not using class information).

4.1 Toy data

We first consider a toy data set, which consists of 80 training points lying on the curve $t_2 = \sin(t_1)$. The coordinate t_1 was generated uniformly in the interval $[-\pi/2, 7\pi/2]$, and spherical Gaussian noise with standard deviation 0.1 was added to the t_2 coordinate. We have defined 4 classes as shown in

Figure 14.1. We also sampled a (complete) test data set of 200 data points from the same distribution.

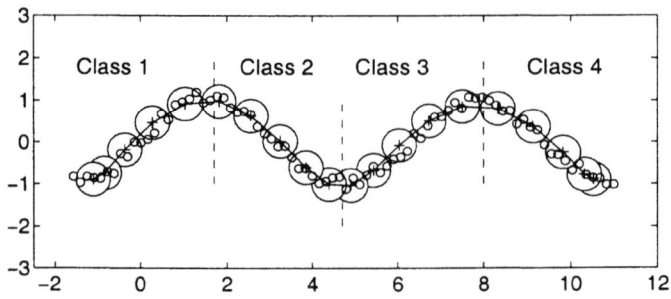

Figure 14.1. GTM at iteration 20 trained on complete data set.

Some values of t_1 were removed at random from points in all classes. A straightforward application of the generic algorithm (III) gave unsatisfactory results because of multi-modalities in the responsibilities (14.2). For example, if $t_2 = 0$, t_1 could be $0, \pi, 2\pi$ or 3π with equal probability. The network was trained using Algorithms I and II. The proportion of missing values was varied from 0 to 70 percent. For the different percentage of missing values, the experiments were carried out 10 times deleting different values.

Plots (a) and (b) in Figure 14.2 display results using Algorithms I and II. It suggests that the EM algorithm combining the class-conditional information is highly successful, while the missing values estimated using class-conditional MI lie on straight lines. Note how the greater uncertainty in estimating missing data for Algorithm II leads to much larger variance in the trained GTM. Plot (c) displays the test set likelihood. It shows that the network can perform better using Algorithm I than using Algorithm II and still performs well with even up to 60% missing values in the training data set.

4.2 Oil Flow Data

In this experiment, we chose an oil pipeline flow data set [1], which is a more realistic test case with 12 measured variables. There are three flow classes in this data set: homogeneous, annular and laminar. A GTM with a two-dimensional latent space was used to model and visualise the data. In the training set, 50% of the data points in each class are incomplete, with between 6 and 9 values removed.

Figure 14.3 displays the results. It suggests that Algorithm I can be an improvement over the two other algorithms for missing data. Plot (b) shows better separation of classes and matches better to the result obtained from the complete data set (plot (a)). After using class-conditional MI, some strongly

Visualisation of Incomplete Data Using Class Information Constraints 171

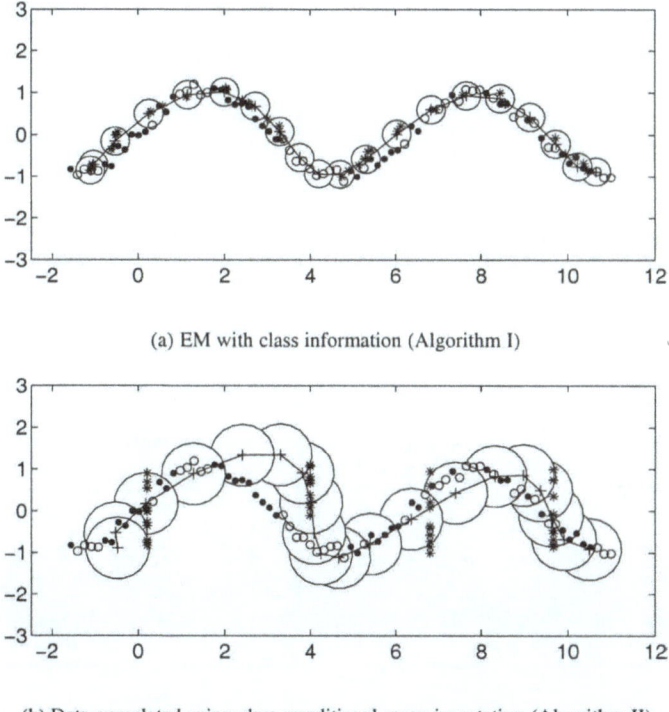

(a) EM with class information (Algorithm I)

(b) Data completed using class-conditional mean imputation (Algorithm II)

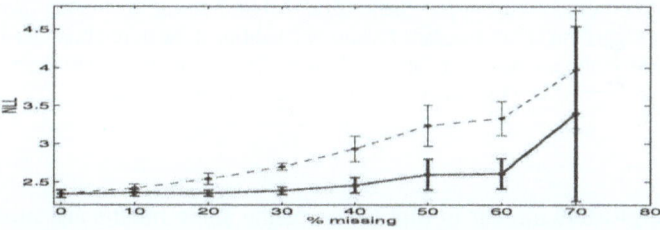

(c) Testing error in the toy problem

Figure 14.2. Toy problem: illustrations of GTMs at iteration 20 ((a)–(b)) trained on data with 50% of t_1 values missing. Complete points are plotted as circles. Stars represent the estimated missing values, while the dots show the original values. The centres of the GMM are plotted as plus signs, and are joined by a line according to their ordering in the (one-dimensional) latent space (K = 20). The discs surrounding each plus sign represent two standard deviations width of the noise model. (c) The average test set negative log likelihood (NLL) and standard deviation error bars for 10 repetitions of the training process. Algorithm I and II are represented by the thick solid and dashed lines, respectively.

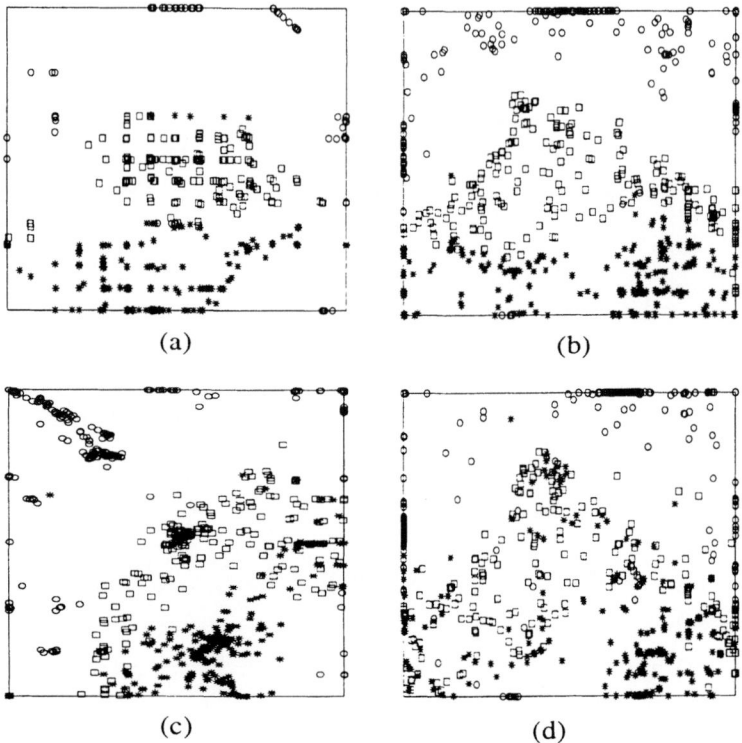

Figure 14.3. Oil flow data: (a) training on the complete data set. The remaining plots show results for models trained on incomplete data. (b) EM algorithm combining the class-conditional information (Algorithm I); (c) standard EM using conditional mean imputation (Algorithm II); (d) generic EM without class information (Algorithm III). The classes homogeneous, annular and laminar are represented by square, star and circle signs respectively.

overlapped clusters appear in plot (c) since the same means are substituted for missing values of the same class. As for plot (d), which was obtained just by the generic algorithm, the homogeneous and annular classes are not separated well as we did not use the class-conditional prior knowledge in the training process.

5. Discussion

In this paper, we have shown how incomplete data and class information can both be included in the GTM training process. The new algorithm is preferable to the simple strategy of just filling in the missing values with conditional means. The problem of multi-modalities in the GTM responsibilities for missing data can be overcome (at least to some degree) by means of the class-conditional prior and this significantly improves both the visualisation plot and the fit of the model.

Acknowledgments

This research has been funded by Pfizer Central Research and BBSRC grant 92/BIO12093. The experiments were carried out with the NETLAB neural network toolbox, available from
http://www.ncrg.aston.ac.uk/netlab.

The authors are most grateful to the anonymous reviewers for their insightful and helpful comments.

References

[1] C. M. Bishop and G. D. James. Analysis of Multi-phase Flows Using Dual-energy Gamma Densitometry and Neural Networks. *Nuclear Instruments and Methods in Physics Research, A*, 327:580–593, 1993.

[2] C. M. Bishop, M. Svensen, and C. K. I. Williams. GTM: The Generative Topographic Mapping. *Neural Computation*, 10(1):215–235, 1998.

[3] C.M. Bishop. *Neural Networks for Pattern Recognition*. Oxford University Press, New York, N.Y., 1995.

[4] A.P. Dempster, N.M. Laird, and D.B. Rubin. Maximum Likelihood from Incomplete Data via the EM Algorithm. *J. Roy. Stat. Soc. B*, 39:1–38, 1977.

[5] Z. Ghahramani and M. I. Jordan. Learning from incomplete data. Technical report, AI Laboratory, MIT, 1994.

[6] Y. Sun, P. Tino, and I. Nabney. GTM-based Data Visualisation with Incomplete Data. Technical Report NCRG/2001/013, NCRG, Aston University, 2001.

[7] Y. Sun, P. Tino, and I. Nabney. Visualisation of incomplete data using class information constraints. Technical Report NCRG/2001/018, NCRG, Aston University, 2001.

[8] V. Tresp, S. Ahmad, and R. Neuneier. Training neural networks with deficient data. In Jack D. Cowan, Gerald Tesauro, and Joshua Alspector, editors, *Advances in Neural Information Processing Systems*, volume 6, pages 128–135. Morgan Kaufmann Publishers, Inc., 1994.

[9] V. Tresp, R. Neuneier, and S. Ahmad. Efficient methods for dealing with missing data in supervised learning. In D. S. Touretzky G. Tesauro and T. K. Leen, editors, *Advances in Neural Information Processing Systems*, volume 7, 1995.

Chapter 15

TOWARDS VIDEOREALISTIC SYNTHETIC VISUAL SPEECH

Barry Theobald, J. Andrew Bangham, Silko Kruse, Gavin Cawley
University of East Anglia, Norwich NR4 7TJ, United Kingdom
b.theobald@uea.ac.uk, {ab,smk,gcc}@sys.uea.ac.uk

Iain Matthews
Robotics Institute, Carnegie Mellon University, Pittsburgh, PA 15213, USA
iainm@cs.cmu.edu

Abstract   In this paper we present preliminary results of work towards a videorealistic visual speech synthesiser. A generative model is used to track the face of a talker uttering a series of training sentences and an inventory of synthesis units is built by representing the trajectory of the model parameters with spline curves. A set of model parameters corresponding to a new utterance is formed by concatenating spline segments corresponding to synthesis units in the inventory and sampling at the original frame rate. The new parameters are applied to the model to create a sequence of images corresponding to the talking face.

Keywords: Shape and appearance models, principal component analysis, visual speech synthesis, facial animation

1. Introduction

Research in computer facial animation began in the early seventies with the pioneering work of Parke [Parke, 1974] and has received increasing interest since [Parke and Waters, 1996]. Applications for facial animation systems are as widespread as surgical planning, low bandwidth video conferencing, media/film production, human-computer interaction, virtual humans in computer games, and psychological experiments in the perception and understanding of speech.

Much of the early work focussed on *computer graphics* based techniques. A popular approach is to represent points on the surface of the face as vertices in a 3D space and approximate the surface itself by connecting the vertices to form a polygonal mesh. Parameters to animate the mesh are either derived empirically through observation [Cohen and Massaro, 1994, Le Goff and Benoit, 1996, Parke, 1974] or are based on some anatomical model [Lee et al., 1993, Platt and Badler, 1981, Waters, 1987]. To make the skin appear more life-like, a facial image can be used as a texture map. However, as the model is animated the static nature of the skin texture is revealed, and even models with very complex animation control parameters do not convince the viewer that they are seeing a human face.

More recent systems have focussed on *image* based techniques in an attempt to achieve video realism. Ezzat and Poggio [Ezzat and Poggio, 1997] used a variation of traditional *key-frame* animation, where new sequences are created by interpolating between key-poses in the image domain. Images were selected with key lip shapes and optical flow used to morph between key-frames. Cosatto and Graf [Cosatto and Graf, 1998] segmented the face and head into several distinct parts and collected a library of triphone samples. New sequences are created by stitching together the appropriate mouth shapes and corresponding facial features based on acoustic triphones and a desired facial expression. Bregler's Video Rewrite system [Bregler et al., 1997] automatically segmented video sequences of a person talking and could reanimate a sequence with different speech by blending images from the training sequence in the desired order.

Image based animation is generally less flexible than model based animation. A model based approach allows new facial movements to be created, or existing expressions to be exaggerated through manipulation of the model parameters. This is difficult to achieve with image based approaches which usually replay existing footage in a new order.

In the following section we describe the construction of a face model capable of producing near photorealistic images of the face. Subsequent sections then discuss how this model is applied to creating near videorealistic synthetic visual speech, where a new visual sequence is synchronised with a previously unseen audio signal.

2. Modelling the Face

Following the notation of Cootes [Cootes et al., 1998], a statistical model of shape, the *point distribution model* (PDM), is trained by hand labeling a set of images and performing a principal component analysis (PCA) on the coordinates of the located landmarks (aligned to remove any pose variation). Any training shape can be approximated using $\mathbf{x} \approx \bar{\mathbf{x}} + \mathbf{P}_s \mathbf{b}_s$, where $\mathbf{P}_s$ is the matrix of the first t_s eigenvectors of the covariance matrix, chosen to describe some percentage, say 95%, of the total variation, and $\mathbf{b}_s$ is a vector of t_s shape parameters.

A texture model is computed by warping the training images so the landmarks in each image lie in the position of the mean landmarks derived from all training images. This normalises the shape of the face, which allows the image to be re-sampled with the same number of pixels in every example and a PCA is computed on the re-sampled pixel values. With such a model any texture can be approximated using $\mathbf{a} \approx \bar{\mathbf{a}} + \mathbf{P}_a \mathbf{b}_a$, where $\bar{\mathbf{a}}$ is the mean shape-free image, $\mathbf{P}_a$ is the matrix of the first t_a eigenvectors of the covariance matrix and $\mathbf{b}_a$ a vector of texture parameters.

Each image is now described by a set of shape parameters and a set of texture parameters, $\mathbf{b}_s$ and $\mathbf{b}_a$ respectively. An appearance model is computed by concatenating the PCA scores for the shape and texture models for each image and performing a third PCA (in $t_s + t_a$ dimensions), where the number of parameters, t_s and t_a, are chosen so that typically 95% of the variance of their respective models is captured. The appearance model is given by Equation 15.1:

$$\mathbf{b} \approx \mathbf{Q}\mathbf{c}, \tag{15.1}$$

where $\mathbf{Q}$ is the matrix of eigenvectors of the covariance matrix and $\mathbf{c}$ a vector of parameters that reflect changes in shape and texture of the face. A large range of realistic images can be synthesised given a set of the first t parameters using Equation 15.2:

$$\mathbf{x} \approx \bar{\mathbf{x}} + \mathbf{P}_s \mathbf{W}_s \mathbf{Q}_s \mathbf{c}, \quad \mathbf{a} \approx \bar{\mathbf{a}} + \mathbf{P}_a \mathbf{Q}_a \mathbf{c}, \tag{15.2}$$

$$\mathbf{Q} = \begin{pmatrix} \mathbf{Q}_s \\ \mathbf{Q}_a \end{pmatrix},$$

where the matrix $\mathbf{W}_s$ takes into account the scaling mismatch between the parameters $\mathbf{b}_s$ (which model Euclidean distance) and $\mathbf{b}_a$ (which model pixel RGB intensity). This is computed as shown in [Cootes et al., 1998] and facial animations are generated by controlling the time trajectory of the vector $\mathbf{c}$. Example faces synthesised by the model are shown in Figure 15.1.

This approach differs from other techniques that have applied PCA to the problem of synthesising visual speech in that Hallgren [Hallgren and Lyberg, 1998] tracked physical markers on the face and applied PCA to the coordinates of the tracked markers. Guiard-Marigny [Guiard-Marigny et al., 1996] animated a geometric model of the lips by representing the contours of the lips with mathematical equations and iteratively predicting one coefficient from others. Brooke and Scott [Brooke and Scott, 1998] performed a PCA on the pixel intensities of the general mouth region and used the principal modes to train a hidden Markov model (HMM) synthesiser. Here PCA is used to construct a compact model of both the shape and the texture of the face. The two

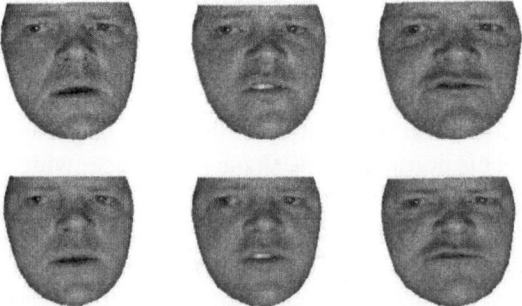

Figure 15.1. The top row shows the face extracted from selected frames of the video and the bottom the synthetic equivalent output by the face tracker.

models are projected into a combined shape and texture space and a concatenation scheme is used to create new synthesised sequences.

3. Data Capture

The training database for the face model was collected using a ELMO EM-02PAL camera and digitized at a frame rate of 25 frames per second using an IEEE 1394 capture card with a frame size of 720x576 (colour). The audio was captured at 11025 Hz stereo and used to phonetically segment the video using an HMM speech recogniser run in forced-alignment mode.

To eliminate unwanted sources of variation from the model, the video was recorded using a head mounted camera, recording a single talker in one sitting, eliminating pose, identity and lighting variations. The speaker held their facial expression as neutral as possible (no emotion) so the variation of the facial features were due to speech.

The training data consisted of 279 sentences containing multiple occurrences of 6315 triphones. The database contained over 30,000 frontal images of the face. A shape model and a texture model were trained on 50 hand-labelled images and the remainder were automatically labelled using the flexible appearance model algorithm [Baker et al., 2001]. The tracker output was manually checked and corrected by hand where necessary.

4. Synthesis

Given the labelled images and the segmented audio, an inventory of synthesis units is constructed. Firstly the shape and texture models used by the tracker are projected into a combined shape and appearance space, as described in section 15.2. Next, the appearance model parameters are computed for each of the training images. Since the landmark positions are known, the shape parameters can be found using $\mathbf{b}_s = \mathbf{P}_s^T(\mathbf{x} - \bar{\mathbf{x}})$ and the texture parameters found using $\mathbf{b}_a = \mathbf{P}_a^T(\mathbf{a} - \bar{\mathbf{a}})$. The shape and texture parameters are concatenated to form a vector $\mathbf{b}$ and the appearance parameters computed using $\mathbf{c} = \mathbf{Q}^T\mathbf{b}$, from Equation 15.1.

The appearance model required 77 parameters to account for 99% of the total variation in the training images and the particular variations captured by the first 3 modes are shown in Figure 15.2.

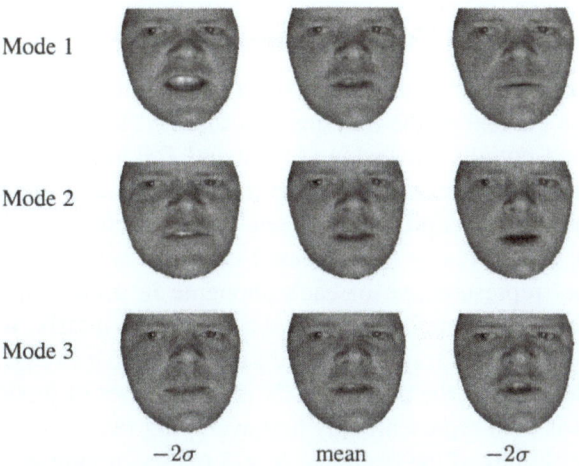

Figure 15.2. The first three modes of variation captured by the combined shape and appearance model at -2, 0 and +2 standard deviations from the mean.

Next, a continuous representation of the trajectories of the original model parameters in each sentence is obtained using Hermite interpolation. A lookup table is created using the speech recogniser output; for each phone the start time in the sentence, the end time, the context and an index indicating in which sentence the phone occurred are stored. The context width is taken as one phone either side of centre phone. Triphones were selected as the synthesis unit as these were the units used by Bregler in the Video Rewrite system [Bregler et al., 1997] and Cosatto in the AT&T visual speech synthesiser [Cosatto and Graf, 1998].

To synthesise a new sentence a sequence of phoneme symbols and durations is required. Presently the system produces a synthetic visual sequence aligned to the audio of a previously unseen utterance. The speech recogniser is used to convert the audio signal to the constituent phoneme symbols and durations. For each phone to be synthesised, the inventory is searched for the closest context to the desired context (see next section). The corresponding portion of the original spline curves are extracted and temporally warped from their original length to the desired length. Note, at present only the portion of the curves corresponding to the centre phones are extracted, the phones either side are used only to determine the context.

The spline segments for each of the synthesis phones are concatenated to form a series of new trajectories, which are sampled at the desired frame rate to give the synthesis model parameters. Smoothing splines [de Boor, 2001] are fitted through the sampled model parameters to ensure a smooth transition from frame to frame and the smoothed parameters are applied to the model (Equation 15.2) to produce the synthetic image sequence of the talking face.

4.1 Synthesising Unseen Contexts

In order to account for unseen contexts, a visual phoneme similarity matrix is used to find a context in the training data that is 'close' to the desired context, as described initially by Arslan [Arslan and Talkin, 1998]. The similarity matrix is automatically derived from the training data and therefore contains an objective measure of similarity between each phoneme and all others.

Arslan built the similarity matrix based on a pair-wise Euclidean distance between vectors representative of each phoneme in a compact shape space. Each occurrence of a phone is represented by an $N \times 5$ matrix, where the original principal component trajectories are sampled at five evenly spaced intervals over the duration of the phone and N is the number of model parameters. Each phoneme is then represented by first averaging the $N \times 5$ matrices, then averaging the columns of the resultant $N \times 5$ matrix, giving an $N \times 1$ vector. The similarity is measured using:

$$\mathbf{S}_{ik} = e^{-\nu \mathbf{D}_{ik}} \qquad (15.3)$$

where $\mathbf{D}_{ik}$ is the distance between phoneme i and k, and ν is a scalar allowing the dynamic range of the similarity to be controlled.

The distance metric used in this work differed to that described above. Each instance of a phone is represented by an $N \times 5$ matrix, as above, and an average $N \times 5$ matrix is computed for each phoneme. The rows of these averaged matrices are weighted according to the significance they have in the appearance model, determined by the percentage of variation captured by the respective mode of variation. The distance between two phonemes is then found by computing the sum of the squared differences between the individual elements of the two matrices and the similarity measured using Equation 15.3.

To find the closest match for an unseen context a measure of similarity is obtained between the desired context and each of the contexts appearing in the inventory using Equation 15.4:

$$\mathbf{s}_j = \sum_{i=1}^{C} \frac{\mathbf{S}_{l_i j}}{i+1} + \sum_{i=1}^{C} \frac{\mathbf{S}_{r_i j}}{i+1} \qquad (15.4)$$

where s_j is the similarity between the desired context and the jth context in the inventory, C is the context width, $\mathbf{S}_{l,j}$ is the similarity between the ith left phoneme in the jth inventory context and the corresponding phoneme in the desired context, $\mathbf{S}_{r,j}$ is the similarity between the ith right phoneme of the jth inventory context and the corresponding phoneme in the desired context. Only the triphones in the inventory with the same centre phone as the desired triphone are searched.

Since triphones are used in this system, C is 1. However, this measure allows the context width of the synthesiser to be easily extended later.

5. Results

The result of applying the synthesis strategy outlined in the previous sections is shown in Figures 15.3 and 15.4. Each of the figures shows the original and synthesised trajectory of the first and second model parameters over the course of a sentence. The first parameter accounted for 27.74% and the second 15.95% of the total variation in the original training image. As an indication of performance the correlation coefficient is used, which is shown in Table 15.1.

Table 15.1. Correlation coefficients for the parameter trajectories shown in Figures 15.3 and 15.4.

	Parameter 1	Parameter 2
Figure 15.3	0.9146	0.8260
Figure 15.4	0.9316	0.7714

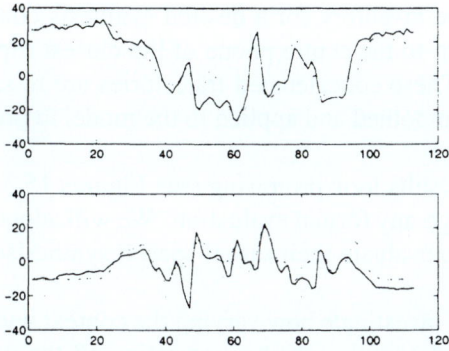

Figure 15.3. Trajectory of the first (top) and second principal component values. The solid black lines are the trajectory of the original parameters and the dashed red lines the trajectory of the synthesiser output. The original data was completely held out from the synthesiser.

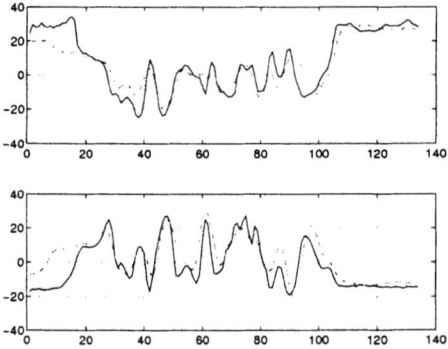

Figure 15.4. Trajectory of the first (top) and second principal component values. The solid black lines are the trajectory of the original parameters and the dashed red lines the trajectory of the synthesiser output. The original data was completely held out from the synthesiser.

6. Summary

In this paper we have described our first generation visual speech synthesiser that is able to produce near videorealistic synthetic visual speech. The basis of the system is a statistical model of the face that produces near photorealistic facial images. The model is used to track the face of a talker in a video sequence, and the trajectories of the tracked model parameters are interpolated using Hermite interpolation. A speech recogniser gives timing information for the triphones appearing in the training sentences and this information is used to construct a synthesis inventory. New sequences are created by locating the best match in the inventory for a desired synthesis context and the spline curves corresponding to the centre phone of the closest triphone are extracted and concatenated. These concatenated trajectories are re-sampled and the resultant parameters smoothed and applied to the model to produce the synthetic image sequence.

While the early results look promising (see Figures 15.3 and 15.4), the system has yet to undergo any formal evaluation. We will adopt the scheme Cohen and Massaro used to evaluate their visual speech synthesiser, Baldi [Massaro, 1998].

Further work will investigate how varying the context width affects the quality of the synthesised speech; whether extracting all the parameters for a triphone and blending overlapping regions gives smoother and more accurate synthetic visual speech; and investigating whether coarticulation models, such as that used by Baldi [Cohen and Massaro, 1994], perform as well as concatenative approaches using the appearance model.

References

[Arslan and Talkin, 1998] Arslan, L. and Talkin, D. (1998). Speech driven 3-d face point trajectory synthesis algorithm. In *Proceedings of the Internation Conference on Speech and Language Processing (ICSLP)*.

[Baker et al., 2001] Baker, S., Dellaert, F., and Matthews, I. (2001). Aligning images incrementally backwards. Technical Report CMU-RI-TR-01-03, Robotics Institute, Carnegie Mellon University, Pittsburgh, PA.

[Bregler et al., 1997] Bregler, C., Covell, M., and Slaney, M. (1997). Video rewrite: driving visual speech with audio. In *Proceedings of SIGGRAPH*, pages 353–360.

[Brooke and Scott, 1998] Brooke, N. and Scott, S. (1998). Two- and three-dimensional audio-visual speech synthesis. In *Proceedings of Auditory-Visual Speech Processing*, pages 213–218.

[Cohen and Massaro, 1994] Cohen, M. and Massaro, D. (1994). Modeling coarticualtion in synthetic visual speech. In Thalmann, N. and D, T., editors, *Models and Techniques in Computer Animation*, pages 141–155. Springer-Verlag.

[Cootes et al., 1998] Cootes, T., Edwards, G., and Taylor, C. (1998). Active appearance models. In Burkhardt, H. and Neumann, B., editors, *Proceedings of the European Conference on Computer Vision*, volume 2, pages 484–498. Springer-Verlag.

[Cosatto and Graf, 1998] Cosatto, E. and Graf, H. (1998). Sample-based synthesis of photo-realistic talking heads. In *Proceedings of Computer Animation*, pages 103–110.

[de Boor, 2001] de Boor, C. (2001). Calculation of the smoothing spline with weighted roughness measure. *Mathematical Models and Methods in Applied Sciences*, 11(1):33–41.

[Ezzat and Poggio, 1997] Ezzat, T. and Poggio, T. (1997). Videorealistic talking faces: A morphing approach. In *Proceedings of the Audiovisual Speech Processing Workshop*, Rhodes, Greece.

[Guiard-Marigny et al., 1996] Guiard-Marigny, T., Tsingos, N., Adjoudani, A., Benoit, C., and Gascuel, M. (1996). 3d models of the lips for realistic speech animation. In *Computer Graphic 96*.

[Hallgren and Lyberg, 1998] Hallgren, A. and Lyberg, B. (1998). Visual speech synthesis with concatenative speech. In *Proceedings of Auditory-Visual Speech Processing*, pages 181–183.

[Le Goff and Benoit, 1996] Le Goff, B. and Benoit, C. (1996). A text-to-audiovisual-speech synthesizer for french. In *Proceedings of the Interna-*

tional Conference on Speech and Language Processing (ICSLP), Philadelphia, USA.

[Lee et al., 1993] Lee, Y., Terzopoulos, D., and Waters, K. (1993). Constructing physics-based facial models of individuals. In *Proceedings of Graphics Interface*, pages 1–8.

[Massaro, 1998] Massaro, D. (1998). *Perceiving Talking Faces*. The MIT Press.

[Parke, 1974] Parke, F. (1974). *A Parametric Model for Human Faces*. PhD thesis, University of Utah, Salt Lake City, Utah, USA.

[Parke and Waters, 1996] Parke, F. and Waters, K. (1996). *Computer Facial Animation*. A K Peters.

[Platt and Badler, 1981] Platt, S. and Badler, N. (1981). Animating facial expression. *Computer Graphics*, 15(3):245–252.

[Waters, 1987] Waters, K. (1987). A muscle model for animating three-dimensional facial expressions. *Proceeding of ACM SIGGRAPH*, 21(4):17–24.

Chapter 16

PROPERTIES OF THE COMPANION MATRIX RESULTANT FOR BERNSTEIN POLYNOMIALS

Joab R. Winkler
Department of Computer Science
The University of Sheffield
Regent Court
211 Portobello Street
Sheffield S1 4DP
United Kingdom
j.winkler@dcs.shef.ac.uk

Abstract   The computational implementation in a floating point environment of the companion matrix resultant is considered and it is shown that the numerical condition of the resultant matrix is strongly dependent on the basis in which the polynomials are expressed. In particular, a companion matrix of a Bernstein polynomial is derived and this is used to construct a resultant matrix for two Bernstein polynomials. A measure of the numerical condition of a resultant matrix is developed and then used to compare the stability of the resultant matrices of the same polynomials that are expressed in different bases. It is shown that it is desirable to express the polynomials in the Bernstein basis, but since the power basis is the natural choice in many applications, a transformation of the resultant matrix between these bases is required. It is shown that this transformation of the resultant matrix between the bases cannot be achieved by performing a basis transformation of each polynomial. Rather, the equation that defines the transformation of the companion matrix resultant between the bases is derived by considering the eigenvectors of the companion matrix of a polynomial in each basis. The numerical condition of this equation is considered and it is shown that it is ill–conditioned, even for polynomials of low degree.

Keywords: Resultants, Bernstein basis, numerical condition

Introduction

The discovery of Groebner bases and the numerous applications of resultants, for example, robot motion planning [1], computer graphics [3], inverse kinematics for serial mechanisms [4], computer vision [6] and computer–aided geometric design [7], have caused renewed interest in computational algebraic geometry. This revival has yielded new theoretical results and generated interest in the computational implementation of resultants, and this paper addresses some issues in the latter topic. In particular, it is shown that the implementation of resultants in a floating point environment poses significant challenges; the polynomial basis may significantly affect the accuracy of computed solution, and advanced methods from computational linear algebra are required in order to estimate the numerical condition of a resultant matrix, and this measure may be used to determine if the computed solution is acceptable.

A resultant of two polynomials is a scalar function of their coefficients that is exactly zero if and only if they have a non–constant common divisor. The resultant is evaluated as the determinant of a matrix, called the resultant matrix, and the rank deficiency of this matrix is equal to the degree of the greatest common divisor (GCD) of the polynomials. Furthermore, the coefficients of the GCD may be evaluated by reducing the resultant matrix to row echelon form. The numerical computation of the resultant of two (or more) polynomials is difficult because of the requirement that the determinant of the resultant matrix be *exactly* zero; since the roots of a polynomial are unaltered by scaling its coefficients by an arbitrary non–zero constant, the resultant matrix and its determinant can also be scaled arbitrarily, and thus a small value of the determinant of this matrix does not imply that there exist one or more roots from each polynomial that are close.

The practical applications of resultants generate entries of the resultant matrix that are specified within a tolerance; they must be interpreted as typical values, and another set of data that lies within the tolerance is also permissible. This family of permissible input data yields a family of solutions, and this leads naturally to the numerical condition of the equation that relates the sets of data and solutions. This concept is well–developed for the standard linear algebraic equation $Ax = b$, such that a computationally reliable solution can be obtained even if this equation is ill–conditioned. Since the solution is a continuous function of the input, the perturbed input $b + \delta b$, which also lies in the family of permissible data, will generate a change δx in the solution. The perturbed solution $x + \delta x$ is acceptable if the equation is well–conditioned.

The situation is more involved when resultants are implemented in a floating point environment because, as noted above, the degree of the GCD is equal to the rank deficiency of the resultant matrix, and the rank of a matrix (an integer quantity) is not a continuous function of the data (which are floating point numbers). Since the rank of a matrix is not defined in a floating point environment, the concepts that are applicable to the linear algebraic equation $Ax = b$ must be discarded or modified. Even if the data are perfect and the polynomials have a non–constant common divisor, roundoff error may cause the *numerical rank* of the resultant matrix to be ill–defined. This problem is compounded when there is an error tolerance on the polynomial coefficients because the family of permissible data may allow the rank of the resultant matrix to take one of several values, depending on the exact perturbation of the polynomial coefficients. Moreover, if the change in the polynomial coefficients is small and causes a reduction in the rank of the resultant matrix, then this matrix is ill–conditioned because it is 'near' a matrix of lower rank.

This paper considers the situation that arises when the data are perfect, and thus consideration is restricted to the effect of roundoff error on the computation of the resultant of two polynomials. It is shown that the numerical condition of the resultant matrix may be significantly improved by representing the polynomials in the Bernstein basis rather than the power basis. There exist several different resultants, including the Sylvester, Bezout and companion matrix resultants, but only the companion matrix resultant is considered in this paper. Some of the results, for example, the expression for the condition number of a resultant matrix, may be applied to the other resultants, and preliminary computational results show that the numerical condition of the Bezout resultant matrix may be significantly improved by expressing the polynomials in the Bernstein basis.

A review of previous work is considered in section 16.1 and the method that is currently used to calculate the resultant of two Bernstein polynomials is described. In particular, it is shown that this method reduces to the calculation of the resultant of two power basis polynomials. Although this technique is theoretically correct, it cannot be justified numerically because one of the advantages of the Bernstein basis – its enhanced numerical stability – is lost. It is therefore advantageous to use the Bernstein basis exclusively, such that the power basis is not utilised. This objective is achieved in section 16.2 because a companion matrix of a Bernstein polynomial is developed and then used to construct a resultant matrix for two Bernstein polynomials.

A comparison of the numerical condition of two resultant matrices requires that an expression for the condition number of a resultant matrix for two polynomials that are expressed in an arbitrary basis be defined. This is considered in section 16.3 and it is shown that this measure is easy to compute and yields a simple geometric interpretation. This measure is used in section 16.4 to compare the numerical condition of the companion matrix resultant C_B of two Bernstein polynomials, and the numerical condition of the resultant matrix C_P that is obtained when a parameter substitution is used to transform, in a weak sense, each Bernstein polynomial to its power basis form. The computational results show that C_B is better conditioned than C_P, possibly by several orders of magnitude, and it may therefore be desirable to transform the power basis resultant matrix to its Bernstein form if the polynomials are specified in the power basis. This topic is considered in section 16.5, and the numerical condition of the equation that defines the transformation is reviewed. A discussion of future work is contained in section 16.6.

1. Previous work

The Bernstein basis is used for the representation of curves and surfaces in computer–aided geometric design because of its elegant geometric properties and simple algorithms that are available for processing it [7], and it is therefore necessary to develop resultants for Bernstein polynomials. This has been achieved hitherto by a simple parameter transformation that enables a Bernstein polynomial to be transformed, in a weak sense, to a power basis polynomial, such that a resultant matrix for two power basis polynomials may be used.

The parameter substitution

$$t = \frac{x}{1-x}, \qquad x \neq 1, \qquad (16.1)$$

transforms the Bernstein polynomial

$$p(x) = \sum_{i=0}^{n} a_i \binom{n}{i} (1-x)^{n-i} x^i,$$

to the power basis polynomial

$$q(t) = (1+t)^n p\left(\frac{t}{1+t}\right) = \sum_{i=0}^{n} c_i t^i, \qquad c_i = a_i \binom{n}{i}, \qquad t \neq -1,$$

and hence the resultant of the Bernstein polynomials $r(x)$ and $s(x)$, with coefficients $\{r_i\}_{i=0}^{n}$ and $\{s_i\}_{i=0}^{m}$, respectively, may be determined by computing the resultant of the power basis polynomials whose coefficients are $\{r_i\binom{n}{i}\}_{i=0}^{n}$ and $\{s_i\binom{m}{i}\}_{i=0}^{m}$. Clearly, if $x = x_0$ is a comon root of the Bernstein polynomials $r(x)$ and $s(x)$, then $t = t_0 = \frac{x_0}{1-x_0}$ is a common root of the power basis polynomials whose coefficients are $\{r_i\binom{n}{i}\}_{i=0}^{n}$ and $\{s_i\binom{m}{i}\}_{i=0}^{m}$. This method may be used for theoretical investigation, but the poor numerical condition of the power basis suggests that the change of polynomial basis that is implied by the substitution (16.1) will yield unsatisfactory results in a floating point environment. It is therefore desirable to develop a resultant matrix for Bernstein polynomials, and thus retain the numerical superiority of the Bernstein basis, and this is considered in the next section.

2. A companion matrix resultant for two Bernstein polynomials

A companion matrix C of a polynomial $p(x)$, expressed in an arbitrary basis, satisfies

$$\det(C - \lambda I) = p(\lambda),$$

where the structure of C is explicitly defined for each basis. Clearly, the eigenvalues of C are identically equal to the roots of $p(x)$. It is shown in this section that a resultant matrix for two Bernstein polynomials can be developed from the companion matrix of one of these polynomials, and the coefficients of the other polynomial. The matrix C assumes a particularly simple form when $p(x)$ is expressed in the power basis, and the following theorem [8] shows that the companion matrix of a Bernstein polynomial has a more complex structure.

THEOREM 16.1 *Consider the matrices A and E, both of order n,*

$$A = \begin{bmatrix} 0 & 1 & 0 & 0 & \cdots & 0 & 0 \\ 0 & 0 & 1 & 0 & \cdots & 0 & 0 \\ \cdot & \cdot & \cdot & \cdot & \cdots & \cdot & \cdot \\ 0 & 0 & 0 & 0 & \cdots & 0 & 1 \\ -a_0 & -a_1 & -a_2 & -a_3 & \cdots & -a_{n-2} & -a_{n-1} \end{bmatrix}, \quad (16.2)$$

and

$$E = \begin{bmatrix} \frac{\binom{n}{1}}{\binom{n}{0}} & 1 & 0 & 0 & \cdots & 0 & 0 \\ 0 & \frac{\binom{n}{2}}{\binom{n}{1}} & 1 & 0 & \cdots & 0 & 0 \\ \cdot & \cdot & \cdot & \cdot & \cdots & \cdot & \cdot \\ 0 & 0 & 0 & 0 & \cdots & \frac{\binom{n}{n-1}}{\binom{n}{n-2}} & 1 \\ -a_0 & -a_1 & -a_2 & -a_3 & \cdots & -a_{n-2} & -a_{n-1} + \frac{\binom{n}{n}}{\binom{n}{n-1}} \end{bmatrix}. \quad (16.3)$$

Then

$$\det(A - \lambda E) = (-1)^n \sum_{i=0}^{n} a_i \binom{n}{i} (1 - \lambda)^{n-i} \lambda^i, \quad a_n = 1,$$

and thus if $\det E \neq 0$, then $M = E^{-1}A$ is a companion matrix of the Bernstein polynomial

$$p(\lambda) = (-1)^n \sum_{i=0}^{n} a_i \binom{n}{i} (1 - \lambda)^{n-i} \lambda^i, \quad a_n = 1.$$

The condition $a_n = 1$ may be generalised to $a_n \neq 0$, and it is therefore assumed that $\lambda_0 = 1$ is not a root of $p(\lambda)$.

It follows from (16.2) and (16.3) that $M = E^{-1}A = (F + A)^{-1}A$ where

$$F = \text{diag}\left[\begin{array}{cccc} \binom{n}{1} & \binom{n}{2} & \cdots & \binom{n}{n-1} & \binom{n}{n} \\ \binom{n}{0} & \binom{n}{1} & & \binom{n}{n-2} & \binom{n}{n-1} \end{array}\right],$$

and a closed form expression for E^{-1} is developed in [8]. This yields

$$M = E^{-1}A = (F + A)^{-1}A = \left(I + \frac{De_n a^T}{\tau}\right) DA,$$

where the elements of D are

$$d_{i,k+i} = \begin{cases} \dfrac{(-1)^k}{\prod_{m=i}^{i+k} c_{mm}} & 1 \leq i \leq n, \quad 0 \leq k \leq n-i, \\ 0 & \text{otherwise,} \end{cases}$$

the elements c_{ij} are

$$c_{ij} = \begin{cases} \dfrac{\binom{n}{i}}{\binom{n}{i-1}} & \text{if } i = j, \\ 1 & \text{if } i + 1 = j, \\ 0 & \text{otherwise,} \end{cases}$$

e_n is the nth standard basis vector, $\tau = \det E$ and

$$a^T = \begin{bmatrix} a_0 & a_1 & \cdots & a_{n-2} & a_{n-1} \end{bmatrix}.$$

The following two theorems show that a companion matrix of a polynomial may be used to construct a resultant matrix for a pair of polynomials.

THEOREM 16.2 *Let $r(x)$ and $s(x)$ be two Bernstein polynomials with coefficients $\{r_j\}_{j=0}^n, r_n = 1$, and $\{s_j\}_{j=0}^m$, respectively,*

$$r(x) = \sum_{j=0}^n r_j \binom{n}{j}(1-x)^{n-j} x^j, \quad s(x) = \sum_{j=0}^m s_j \binom{m}{j}(1-x)^{m-j} x^j. \tag{16.4}$$

If M is a companion matrix of the polynomial $r(x)$ and the eigenvalues of M are $\{\lambda_i\}_{i=1}^n$, then

$$\det(s(M)) = \prod_{i=1}^n s(\lambda_i), \tag{16.5}$$

and thus the determinant of $s(M)$ is equal to zero if and only if λ_i is a root of $s(x)$. Since the eigenvalues $\{\lambda_i\}_{i=1}^n$ are the roots of $r(x)$, it follows that $s(M)$ is a resultant matrix for the polynomials $r(x)$ and $s(x)$.

Proof Consider the matrix polynomial

$$s(M) = \sum_{j=0}^{m} s_j \binom{m}{j} (I - M)^{m-j} M^j.$$

If the eigenpairs of M are $\{\lambda_i, x_i\}_{i=1}^n$, the eigenpairs of

$$\binom{m}{j} (I - M)^{m-j} M^j,$$

are

$$\left\{ \binom{m}{j} (1 - \lambda_i)^{m-j} \lambda_i^j, x_i \right\}_{i=1}^n,$$

and thus

$$\binom{m}{j} (I - M)^{m-j} M^j x_i = \binom{m}{j} (1 - \lambda_i)^{m-j} \lambda_i^j x_i, \qquad i = 1, \ldots, n.$$

It follows that for $i = 1, \ldots, n$,

$$\sum_{j=0}^{m} s_j \binom{m}{j} (I - M)^{m-j} M^j x_i = \sum_{j=0}^{m} s_j \binom{m}{j} (1 - \lambda_i)^{m-j} \lambda_i^j x_i,$$

and thus the eigenvalues of $s(M)$ are

$$s(\lambda_i) = \sum_{j=0}^{m} s_j \binom{m}{j} (1 - \lambda_i)^{m-j} \lambda_i^j, \qquad i = 1, \ldots, n,$$

and hence (16.5) is obtained. □

The following theorem enables the degree and coefficients of the GCD of $r(x)$ and $s(x)$ to be calculated [2, 5].

THEOREM 16.3 *Let $w(x)$ be the GCD of $r(x)$ and $s(x)$, which are defined in (16.4), and let M be the companion matrix of $r(x)$. Then*

1 *The degree of $w(x)$ is equal to $n - \text{rank}\, s(M)$.*

2 *The coefficients of $w(x)$ are proportional to the last row of $s(M)$ after it has been reduced to row echelon form.*

3. A condition number of a resultant matrix

A normwise condition number of the resultant matrix $s(M)$ is developed and it is shown that it is independent of the arbitrary scale factor that can be applied to the polynomial $s(x)$. An expression for this condition number is stated in the following theorem. The result has been specialised to square matrices because a resultant matrix is always of this form.

THEOREM 16.4 *Let $X \in \mathbb{R}^{n \times n}$ be of rank r, and let USV^T be its singular value decomposition. Let $X_k = US_k V^T$ where $k < r$, and let $S_k \in \mathbb{R}^{n \times n}$ be the diagonal matrix with elements*

$$\begin{pmatrix} \sigma_1 & \sigma_2 & \cdots & \sigma_k & 0 & 0 & \cdots & 0 \end{pmatrix}, \qquad \sigma_1 \geq \sigma_2 \geq \cdots \geq \sigma_k > 0.$$

Then the rank of X_k is k, and

$$\sigma_{k+1} = \|X - X_k\|_2 = \min_{rank\,(Y)=k} \|X - Y\|_2.$$

This theorem states that if a square matrix X is of order n and rank $r \leq n$, then the nearest matrix Y, of rank $k < r$, to X is obtained by setting $\sigma_i = 0, i = k+1, \ldots, r$, where σ_i is the ith singular value of X. Furthermore, the minimum normwise distance between X and Y is σ_{k+1}.

This expression for the minimum normwise distance between a matrix X and the nearest matrix Y of specified lower rank cannot be used to determine the numerical condition of $s(M)$ because it is not scale–invariant. In particular, the polynomial $r(x)$ is monic and therefore normalised, but the polynomial $s(x)$ can be scaled arbitrarily. It follows that if $s(x)$ is scaled to $\alpha s(x)$, then the singular values of $s(M)$ are also scaled by α, and thus the minimum normwise distance between $s(M)$ and the nearest matrix of lower rank is also scaled by α. This is unsatisfactory, but this deficiency is easily overcome by defining a scale invariant measure, called the *normalised distance to singularity*. The normalised distance to singularity between the matrices X and X_k that are defined in theorem 16.4 is obtained by normalising σ_{k+1} by $\|X\|_2 = \sigma_1$,

$$\frac{\sigma_{k+1}}{\sigma_1}, \qquad (16.6)$$

which may be defined as the reciprocal of the condition number of $X_{k+1} = US_{k+1}V^T$.

The application of theorem 16.4 to the determination of the numerical condition of a resultant matrix requires that a unit loss of rank be considered, and thus $k = r - 1$. It follows from (16.6) that if $s(M)$ has rank r, the normalised distance to singularity of $s(M)$ is

$$d(s(M)) = \frac{\sigma_r(s(M))}{\sigma_1(s(M))}. \qquad (16.7)$$

The parameter substitution (16.1) may be used to transform (in a weak sense) a Bernstein polynomial to the power basis, and its application to the polynomials (16.4) yields the polynomials

$$\tilde{r}(x) = \sum_{j=0}^{n} \tilde{r}_j x^j, \qquad \tilde{r}_j = r_j \binom{n}{j},$$

and

$$\tilde{s}(x) = \sum_{i=0}^{m} \tilde{s}_i x^i, \qquad \tilde{s}_i = s_i \binom{m}{i},$$

respectively. If P is a companion matrix of $\tilde{r}(x)$, then

$$\tilde{s}(P) = \sum_{i=0}^{m} \tilde{s}_i P^i,$$

is a resultant matrix for $\tilde{s}(x)$ and $\tilde{r}(x)$, and it follows from (16.7) and its equivalent for $\tilde{s}(P)$ that the ratio of the normalised distance to singularity of $\tilde{s}(P)$ to the normalised distance to singularity of $s(M)$ is

$$d(\tilde{s}(P), s(M)) = \frac{\sigma_r(\tilde{s}(P))}{\sigma_1(\tilde{s}(P))} \frac{\sigma_1(s(M))}{\sigma_r(s(M))}. \tag{16.8}$$

This measure is used in section 16.4 to compare the normalised distance to singularity of $\tilde{s}(P)$ and $s(M)$.

4. Computational experiments

This section contains several examples that enable a comparison to be made of the resultant matrix of the Bernstein polynomials $r(x)$ and $s(x)$, and the power basis resultant matrix that is obtained when the substitution (16.1) is used to transform $r(x)$ and $s(x)$ to $\tilde{r}(x)$ and $\tilde{s}(x)$, respectively. The ratio (16.8) is used to calculate the ratio of the normalised distance to singularity of $\tilde{s}(P)$ and $s(M)$. Since use of the Bernstein basis necessarily implies that interest is restricted to the unit interval $I = \{x : 0 \le x \le 1\}$, the polynomials $r(x)$ and $s(x)$ are chosen so that they have at least one common root in this interval.

In all the examples, the polynomial $r(x)$ is defined as the Wilkinson polynomial,

$$r(x) = \prod_{i=1}^{19} \left(x - \frac{i}{20}\right) = \sum_{i=0}^{19} r_i \binom{19}{i} (1-x)^{19-i} x^i, \tag{16.9}$$

and different polynomials $s_i(x)$ are used. The upper index is 19 and not its more usual value of 20 because, as noted in theorem 16.1, the companion matrix M must not have an eigenvalue $\lambda_0 = 1$. The matrices $s_i(M)$ and $\tilde{s}_i(P)$

were computed for several polynomials $s_i(x)$, and the ratio of the normalised distances to singularity $d(\tilde{s}_i(P), s_i(M))$ was calculated. The results are shown in table 16.1.

Table 16.1. The ratio $d(\tilde{s}_i(P), s_i(M))$ for the polynomials (16.9) and $s_i(x)$.

Experiment	Polynomial $s_i(x)$	$d(\tilde{s}_i(P), s_i(M))$
1	$s_1(x)$	2.728×10^{-2}
2	$s_2(x)$	2.194×10^{-15}
3	$s_3(x)$	7.038×10^{-13}
4	$s_4(x)$	1.828×10^{-4}
5	$s_5(x)$	1.434×10^{-13}
6	$s_6(x)$	1.747×10^{-12}
7	$s_7(x)$	2.235×10^{-12}
8	$s_8(x)$	2.094×10^{-9}
9	$s_9(x)$	8.449×10^{-1}
10	$s_{10}(x)$	6.785×10^{1}

The polynomials $s_i(x), i = 1, \ldots, 10$, in this table are :

$$
\begin{aligned}
s_1(x) &= (x - 0.80)(x - 0.90)(x - 0.95)(x - 0.99)(x - 1.01) \\
s_2(x) &= (x + 0.05)(x + 0.025)(x - 0.05)^3 \\
s_3(x) &= (x - 0.20)^2 (x - 0.40)^2 (x - 0.60)^2 (x - 0.79)(x - 0.80) \\
s_4(x) &= (x + 2.0)(x + 1.0)(x - 0.30)(x - 0.50)(x - 0.70)(x - 0.80) \times \\
& \quad (x - 0.85)(x - 0.86)(x - 0.87)(x - 0.95)^2 (x - 1.01) \\
s_5(x) &= (x - 0.02)(x - 0.04)(x - 0.06)(x - 0.08)(x - 0.10)^2 \\
s_6(x) &= (x - 0.05)^3 (x - 0.10)^3 (x - 0.15)^3 \\
s_7(x) &= (x - 0.40)^2 (x - 0.425)(x - 0.45)^2 (x - 0.475)^2 (x - 0.50)^3 \\
s_8(x) &= (x + 3.0)(x + 1.0)(x - 0.40)^2 (x - 0.41)^2 (x - 0.94) \times \\
& \quad (x - 0.95)^2 (x - 0.96)(x - 1.50)(x - 1.60) \\
s_9(x) &= (x - 0.50)(x - 0.55)(x - 0.60)(x - 0.65)(x - 0.70) \times \\
& \quad (x - 0.75)(x - 0.80)(x - 0.85)(x - 0.90)(x - 0.95) \\
s_{10}(x) &= (x - 0.95)^2 (x - 0.97)^2 (x - 0.99)^2 (x - 1.01)^2
\end{aligned}
$$

The results in table 16.1 show that the Bernstein basis resultant matrix $s(M)$ is, in most situations, better conditioned than the power basis resultant matrix $\tilde{s}(P)$ because it is further away from singularity, but the ratio $d(\tilde{s}_{10}(P), s_{10}(M))$ shows that $s(M)$ is not guaranteed to be better conditioned than $\tilde{s}(P)$. More computational results are in [8], and these two sets of results show that, in general, the parameter substitution (16.1) is not recommended because it causes the resultant matrix to be ill–conditioned. It must be noted that these results do not define an exact comparison between the power and Bernstein bases because this parameter substitution is not a basis transformation in the strict sense. A computational experiment that compares the numerical condition of the power and Bernstein resultant matrices when the basis transformation is performed exactly is in [8], and it is noted that the Bernstein basis resultant matrix is much better conditioned.

5. The basis transformation of the companion matrix resultant

The results of section 16.4 show that it is numerically advantageous to compute the resultant of two polynomials when they are expressed in the Bernstein basis rather than when they are expressed in the power basis. The power basis is the natural choice for the representation of curves and surfaces in the applications (apart from computer graphics) that are mentioned in the introduction, and the computation of the resultant of two polynomials in a floating point environment therefore requires that a basis transformation of the resultant matrix be performed.

Consider two power basis polynomials $f(x)$ and $g(x)$, and let $r(x)$ and $s(x)$ be, respectively, their Bernstein forms,

$$r(x) = \sum_{i=0}^{n} r_i \binom{n}{i} (1-x)^{n-i} x^i, \qquad f(x) = \sum_{i=0}^{n} f_i x^i,$$

where $r(x) \equiv f(x)$, and

$$s(x) = \sum_{i=0}^{m} s_i \binom{m}{i} (1-x)^{m-i} x^i, \qquad g(x) = \sum_{i=0}^{m} g_i x^i,$$

where $s(x) \equiv g(x)$. It is important to note that if $f(x)$ and $g(x)$ are given, the resultant of $r(x)$ and $s(x)$ cannot be computed in a floating point environment by performing a basis transformation on $f(x)$ and $g(x)$ and then using the theory in section 16.2 to compute $s(M)$ where M is the companion matrix of $r(x)$. This method decouples the polynomials – the pairs of polynomials $(r(x), f(x))$ and $(s(x), g(x))$ are treated independently, but the resultant matrix $s(M)$ is a function of r_i and s_j, and thus the error that arises from the cross product terms $r_i s_j$ in $s(M)$ must be considered for a complete error analysis.

Let P be the companion matrix of the power basis polynomial $f(x)$, such that it has the same form as A in (16.2). An identical proof to that of theorem 16.2 shows that $g(P)$ is a companion matrix resultant for the power basis polynomials $f(x)$ and $g(x)$. It is shown in [9] that the resultant matrices $s(M)$ and $g(P)$ are related by the similarity transformation

$$g(P) = Bs(M)B^{-1}, \qquad (16.10)$$

where the entries $\{b_{jk}\}_{j,k=0}^{n-1}$ of B and $\{b_{jk}^{-1}\}_{j,k=0}^{n-1}$ of B^{-1} are given by, respectively,

$$b_{jk} = \begin{cases} \frac{n-k}{n} \frac{\binom{k}{j}}{\binom{n-1}{j}}, & j = 0, \ldots, n-1; \; k = j, \ldots, n-1, \\ 0, & k < j, \end{cases} \quad (16.11)$$

and

$$b_{jk}^{-1} = \begin{cases} (-1)^{k-j} \binom{n}{j} \binom{n-1-j}{k-j}, & j = 0, \ldots, n-1; \; k = j, \ldots, n-1, \\ 0, & k < j. \end{cases}$$
(16.12)

The eigenvalues of P are equal to the eigenvalues of M since these are companion matrices of the same polynomial but expressed in different bases. The transformation (16.10) is obtained by deriving an expression for the eigenvectors $y_i, i = 1, \ldots, n$, and $x_i, i = 1, \ldots, n$, of, respectively, P and M, and noting that the elements of each of the eigenvectors y_j and x_j form a basis for polynomials of degree $n - 1$. It follows, therefore, that there exists a square non–singular matrix B such that $y_j = Bx_j$, and this equation allows the closed form expression (16.11) to be derived. It is shown in [9] that B is a totally non–negative matrix, and BB^T and B^TB are oscillation matrices.

The motivation for the transformation (16.10) of the power basis resultant matrix $g(P)$ to its Bernstein form $s(M)$ is the improvement in the numerical condition, as noted in the examples in section 16.4. This advantage of performing all computations in the Bernstein basis is retained only if (16.10) is well–conditioned, such that errors due to floating point computations and uncertainties in the data are not magnified during the computation of $s(M)$ from $g(P)$. If, however, this equation is ill–conditioned, then the advantages of transforming to the Bernstein basis may be cancelled by the potentially large magnification of these errors. It is therefore necessary to determine the numerical condition of (16.10), and this is considered in detail in [9]. It is shown that this equation can be reduced to a linear algebraic equation by the Kronecker product, and this allows a sharp upper bound of the condition number of this equation to be obtained.[1] It is shown in [9] that this condition number increases rapidly with n, and thus the equation may be ill–conditioned, even for matrices B of low order. It is therefore concluded that the resultant matrix should be formed *a priori* in the Bernstein basis, such that the power basis is not used.

The Bernstein basis is the common choice for the representation of polynomials in computer graphics and computer–aided geometric design, but the

---

[1] The author wishes to thank Professor N. Higham of Manchester University for introducing him to the Kronecker product.

power basis is commonly used in other applications. The results in [8] and [9] show that improved numerical answers will be obtained if the Bernstein basis were also the natural choice for the representation of polynomials in these other applications.

6. Discussion

This paper has considered some issues that arise during the computational implementation of resultants in a floating point environment. It was shown that it is desirable to compute the resultant of two polynomials that are expressed in the Bernstein basis rather than perform a parameter substitution that implements a basis transformation, in a weak sense, to the power basis. Attention was restricted to the companion matrix resultant, and it is therefore desirable to construct other resultant matrices for the Bernstein basis and determine if they are also numerically better conditioned than their power basis equivalents. Also, the numerical condition of the Bernstein form of these types of resultant matrix must be compared in order that the most stable resultant matrix be used.

The results that were developed in this paper for the companion matrix resultant are also applicable to other resultant matrices. For example, the normalised distance to singularity (16.7) can be applied to the Bezout resultant matrix, but this measure is not scale invariant for the Sylvester resultant matrix because the coefficients of the polynomials are decoupled in this matrix [10]. Also, the basis transformation equation (16.10) generalises to $S = XZY$ for the Sylvester and Bezout resultant matrices, where S and Z are the resultant matrices in the power and Bernstein bases, and X and Y are transformation matrices whose elements are given by formulae that are similar to those in (16.11) and (16.12).

References

[1] J. F. Canny. *The Complexity of Robot Motion Planning*. The MIT Press, Cambridge, USA, 1988.

[2] R. N. Goldman, T. W. Sederberg, and D. C. Anderson. Vector elimination : A technique for the implicitization, inversion and intersection of planar parametric rational polynomial curves. *Computer Aided Geometric Design*, 1:327–356, 1984.

[3] J. T. Kajiya. Ray tracing parametric patches. *Computer Graphics*, 16:245–254, 1982.

[4] D. Manocha. Numerical methods for solving polynomial equations. In D. Cox and B. Sturmfels, editors, *Proceedings of Symposia in Applied Mathematics, volume 53*, Applications of Computational Algebraic Geometry, pages 41–66. AMS, Rhode Island, USA, 1998.

[5] Y. De Montaudouin and W. Tiller. The Cayley method in computer aided geometric design. *Computer Aided Geometric Design*, 1:309–326, 1984.

[6] S. Petitjean. Algebraic geometry and computer vision : Polynomial systems, real and complex roots. *Journal of Mathematical Imaging and Vision*, 10:191–220, 1999.

[7] T. Sederberg. Applications to computer aided geometric design. In D. Cox and B. Sturmfels, editors, *Proceedings of Symposia in Applied Mathematics, volume 53*, Applications of Computational Algebraic Geometry, pages 67–89. AMS, Rhode Island, USA, 1998.

[8] J. R. Winkler. A companion matrix resultant for Bernstein polynomials, 2002. Submitted.

[9] J. R. Winkler. The transformation of the companion matrix resultant between the power and Bernstein polynomial bases, 2002. Submitted.

[10] J. R. Winkler and R. N. Goldman. The Sylvester resultant matrix for Bernstein polynomials, 2002. Submitted.

Chapter 17

COMPUTATION WITH A NUMBER OF NEURONS

Si Wu, Danmei Chen
Department of Computer Science
Sheffield University
Sheffield S1 4DP
United Kingdom
{s.wu, d.chen}@dcs.shef.ac.uk

Abstract An essential feature of neural information processing in the brain is that a stimulus is not represented by the activity of a single neuron but rather by the joint activities of a number of them. Such a coding strategy is called population coding. This paper reviews the recent progress on the understanding of computational properties of population codes, with emphasis on how to implement a hierarchical Bayesian decoding procedure in a neural circuit.

Keywords: Population coding, Bayesian inference, recurrent network, line attractor, Hebbian learning

1. Introduction

Nowadays we have seen great achievements in the speed of digital computers, and witnessed machines outperform human beings in computational tasks that can be programmed serially. However, there are still many jobs for which the human being remains superior to conventional methods, despite its slowness of signal processing. These include many ordinary life tasks, such as memory retrieving, object recognition and body motion control. The superiority of brain functions is attributed to the essential feature, that is, computation is done parallely by a large number of neurons.

The parallel processing of brain functions is evident by the way of how information is represented in the brain. Experimental study has revealed that a stimulus is not encoded by a single neuron but rather by a population of them. Such a coding strategy is called population coding [3, 5]. For example, the moving direction of an object is represented by the joint activities of a number of neurons in the visual area MT [4]. Apart from being robust to the fluctuation in a single neuron's activity, population coding turns out to have many other computationally desirable properties, and is the unified platform above which the brain completes different calculations (similar to the basis function expansion in nonlinear function approximation) [7]. Understanding population coding is critical for the understanding of high-level brain functions.

In this paper, we will review some recent results on the understanding of computational properties of population codes. The emphasis is on introducing a hierarchical Bayesian decoding procedure and how to implement it in a neural circuit [12].

2. Population Coding Paradigm

The theoretic study of population coding in the literature is often based on a simplified paradigm of the real neural information process, which captures four essential features of neural coding :

- A stimulus is encoded by a number of neurons;
- Neurons' activities fluctuate;
- The mean activity of a neuron is described by the tuning function;
- Neurons' activities are correlated.

This simplified model neglects details possibly associated with particular brain functions, and aims to answer general performances of population codes. It plays the similar role as the Ising model in statistical physics.

We can formulate the above paradigm into a mathematic problem. Consider N neurons coding for a stimulus x. The population activity is denoted by $\mathbf{r} = \{r_i\}$. Here r_i is the response of the ith neuron, which is given by

$$r_i = f_i(x) + \epsilon_i, \qquad (17.1)$$

where $f_i(x)$ is the tuning function representing the mean activity.

The random variable ϵ_i represents the fluctuation. The neural correlation is described by the covariance matrix $\langle \epsilon_i \epsilon_j \rangle$.

The encoding process of a population code is specified by the conditional probability $Q(\mathbf{r}|x)$ (i.e., the model of ϵ_i). The decoding is to infer the value of x from the observed $\mathbf{r}$.

3. Bayesian Population Decoding

Let us consider a general Bayesian decoding method, which estimates the stimulus by maximizing a log posterior distribution, $\ln P(x|\mathbf{r})$, i.e.,

$$\begin{aligned} \hat{x} &= \operatorname{argmax}_x \ \ln P(x|\mathbf{r}), \\ &= \operatorname{argmax}_x \ \ln P(\mathbf{r}|x) + \ln P(x), \end{aligned} \qquad (17.2)$$

where $P(\mathbf{r}|x)$ is the likelihood function. It can be equal to or different from the real encoding model $Q(\mathbf{r}|x)$, depending on the available information of the encoding process. If $P(\mathbf{r}|x) \neq Q(\mathbf{r}|x)$, it is called unfaithful decoding [8]. As shown in [8, 9, 10], there normally exists an optimal unfaithful model, based on which the decoding is a good balance between computational complexity and decoding accuracy.

$P(x)$ is the distribution of x, representing the prior knowledge. The above method is also called Maximum a Posteriori (MAP). When the distribution of x is, or is assumed to be (when there is no prior knowledge) uniform, MAP is equivalent to Maximum Likelihood Inference (MLI).

MAP could be used in the information processing of the brain in several occasions. Let us consider the following scenario : A stimulus is decoded in multiple steps. This happens when the same stimulus is presented through multiple steps, or during a single presentation, neural signals are sampled many times. In both cases, the brain successively gains a rough estimation of the stimulus in each step decoding, which can serve to be the prior knowledge when further decoding is concerned. It is therefore natural to use MAP in this situation. Experiencing slightly different stimuli in consecutive steps as studied in [13], or more generally, stimulus slowly changes with time (multiple-step diagram is a discrete approximation), is a similar scenario. For simplicity, we only consider the case that stimulus is unchanged in the present study.

We should note that the estimation expressed by eq.(17.2) is very general. Indeed, most decoding methods in the literature can be cast in this form with suitable likelihood function and prior distribution. For example, the conventional method, Center of Mass, can be formulated as MLI based on an unfaithful model which neglects neural correlation and uses a tuning function of the quadratic form [11, 13].

4. Network Implementation of Bayesian Decoding

We investigate how to implement MAP by a recurrent network. Two-step decoding is considered for illustration, though generalization to multiple-step is straightforward. In the first step, ML is used when there is no prior knowledge, but MAP is used in the second step.

For clearance, some notations are introduced first. Denote $\hat{x}_t$ a particular estimation of the stimulus in the tth step, and Ω_t^2 the corresponding variance. The prior distribution of x in the $(t+1)$th step is assumed to be a Gaussian with the mean value $\hat{x}_t$, i.e.,

$$P(x|\hat{x}_t) = \frac{1}{\sqrt{2\pi}\tau_t} \exp^{-(x-\hat{x}_t)^2/2\tau_t^2}, \qquad (17.3)$$

where the parameter τ_t reflects the estimator's confidence on $\hat{x}_t$. The posterior distribution of x in the $(t+1)$th step is given by

$$P(x|\mathbf{r}) = \frac{P(\mathbf{r}|x)P(x|\hat{x}_t)}{P(\mathbf{r})}. \qquad (17.4)$$

The solution of MAP is obtained by solving

$$\begin{aligned} \nabla \ln P(\hat{x}_{t+1}|\mathbf{r}) &= \nabla \ln P(\mathbf{r}|\hat{x}_{t+1}) - (\hat{x}_{t+1} - \hat{x}_t)/\tau_t^2, \\ &= 0. \end{aligned} \qquad (17.5)$$

We talk about using a recurrent network to implement MAP in the sense that, the network gives the same solution as eq.(17.5). The network we consider is a fully connected one-dimensional homogeneous neural field (i.e. the continuous extension of the discrete case) [1], in which c denotes the position coordinate, i.e., the neurons' preferred stimuli. The tuning function of the neuron with preferred stimulus c is

$$f_c(x) = \frac{1}{\sqrt{2\pi}a} \exp^{-(c-x)^2/2a^2}. \qquad (17.6)$$

For simplicity, we consider an encoding process in which the fluctuations in neurons' responses are independent Gaussian noises (more general correlated cases can be handled similarly), that is,

$$Q(\mathbf{r}|x) = \frac{1}{Z} \exp^{-\frac{\rho}{2\sigma^2} \int (r_c - f_c(x))^2 dc}, \qquad (17.7)$$

where ρ is the neuron density and Z is the normalization factor. A faithful model is used in both steps decoding, i.e., $P(\mathbf{r}|x) = Q(\mathbf{r}|x)$ (again, generalization to more general cases of $P(\mathbf{r}|x) \neq Q(\mathbf{r}|x)$ is straightforward).

For the above model setting, the solution of MLI in the first step is calculated to be

$$\hat{x}_1 = \mathrm{argmax}_x \int r_c f_c(x) dc, \qquad (17.8)$$

where the condition $\int f_c^2(x) dc = $ constant has been used.

From the above equation, we see that MLI can be intuitively understood as a template-matching process. The template is the tuning function, and the matching is to adjust the position of template through moving the value of x, such that the overlap between the population activity and the template is the largest.

The solution of MAP in the second step is

$$\hat{x}_2 = \mathrm{argmax}_x \left[\int r_c f_c(x) dc - (x - \hat{x}_1)^2 / 2\tau_1^2 \right]. \qquad (17.9)$$

Compared with eq.(17.8), (17.9) has one more term corresponding to the contribution of the prior distribution.

Now we come to the study of using a recurrent network to realize eqs.(17.8) and (17.9). According to the above picture of template-matching, the network must be designed to have three properties:

1. The steady state of the network must have the same shape of tuning function, in order to generate the correct template (the peak position of steady state is the network estimation);

2. When no stimulus exists, the steady state of network must be neutrally stable (and just stable) on a one dimensional attractor (called line attractor) parameterized by all possible stimulus values. This allows the network to decode the whole range of stimulus;

3. An external input (representing stimulus) drifts the network to the position corresponding to the maximum overlap between the population activity and the template.

The above idea was first proposed by Pouget and co-authors [6, 2], and further confirmed by Wu et al.[11, 12]. It builds up the connection between abstract mathematical decoding methods expressed by eq.(17.2) and the reading-out process possibly used by the brain.

Following the three requirements, we construct the following network dynamics. Let U_c denote the (average) internal state of neuron at c, and $W_{c,c'}$ the recurrent connection weights from neurons at c to those at c'. The dynamics of neural excitation is governed by

$$\frac{dU_c}{dt} = -U_c + \int W_{c,c'} O_{c'} dc' + I_c, \qquad (17.10)$$

where

$$O_c = \frac{U_c^2}{1 + \mu \int U_c^2 dc}, \qquad (17.11)$$

is the activity of neurons at c and I_c is the external input arriving at c.

The recurrent interactions are chosen to be

$$W_{c,c'} = \exp^{-(c-c')^2/2a^2}, \qquad (17.12)$$

which ensures that when there is no external input ($I_c = 0$), the network is neutrally stable on line attractor,

$$\tilde{O}_c(z) = D \exp^{-(c-z)^2/2a^2} \quad \forall z, \qquad (17.13)$$

where the parameter D is constant and can be determined easily. Note that the line attractor has the same shape as the tuning function. This is crucial, which allows the network perform template-matching by using the tuning function, being as same as MLI and MAP.

When a sufficiently small input I_c is added, the network is no longer neutrally stable on the line attractor. It can be proved that the steady state of the network has approximately the same shape as eq.(17.13) (the deviation is of the 2nd order of the magnitude of I_c), whereas its steady position on the line attractor (i.e., the network estimation) is determined by maximizing the overlap between I_c and $\tilde{O}_c(z)$ [6, 11].

Thus, if $I_c = \varepsilon r_c$ in the first step, where ε is a sufficiently small number, the network estimation is given by

$$\hat{z}_1 = \text{argmax}_z \int r_c \tilde{O}_c(z) dc, \qquad (17.14)$$

which has the same value as the solution of ML (see eq.(17.8)). We say that the network implements MLI.

To implement MAP in the second step, it is critical to identify a neural mechanism which can 'transmit' the prior knowledge obtained in the first step to the second one. We find that this is naturally done by Hebbian learning.

After the first step decoding, the recurrent interaction changes a small amount according to the Hebbian rule, whose new value is

$$\tilde{W}(c,c') = W_{c,c'} + \eta \tilde{O}_c(\hat{z}_1) \tilde{O}_{c'}(\hat{z}_1), \qquad (17.15)$$

where η is a small positive number representing the Hebbian learning rate, and $\tilde{O}_c(\hat{z}_1)$ is the neuron activity in the first step.

With the new recurrent interactions, the net input from other neurons to the one at c is calculated to be

$$\int \tilde{W}_{c,c'} O_{c'} dc' = \int W_{c,c'} O_{c'} dc' + \eta \tilde{O}_c(\hat{z}_1) \int O_{c'}(\hat{z}_1) O_{c'} dc',$$

$$\approx \int W_{c,c'} O_{c'} dc' + \nu \tilde{O}_c(\hat{z}_1), \qquad (17.16)$$

where ν is a small constant.

Substituting eq.(17.16) in (17.10), we see that the network dynamics in the second step, when compared with the first one, is in effect to modify the input I_c to be $I'_c = \varepsilon(r_c + A\tilde{O}_c(\hat{z}_1))$.

Thus, the network estimation in the second step is determined by maximizing the overlap between I'_c and $\tilde{O}_c(z)$, which gives

$$\hat{z}_2 = \text{argmax}_z \left[\int r_c \tilde{O}_c(z) dc + A \int \tilde{O}_c(\hat{z}_1) \tilde{O}_c(z) dc \right],$$

$$\approx \text{argmax}_z \left[\int r_c \tilde{O}_c(z) dc - B(z - \hat{z}_1)^2 / 4a^2 \right], \qquad (17.17)$$

where B is a positive constant related to the Hebbian learning rate.

Compare eqs.(17.9) and (17.17), we see that the second term in (17.9) plays the same role as the prior knowledge in MAP. Thus, the network indeed implements MAP. The above procedure can be extended to a hierarchical procedure, in which the estimation in the previous step forms the Gaussian prior for the current decoding.

The above theoretic analysis is confirmed by the simulation experiment, as shown in Fig.1, which was done with 101 neurons uniformly distributed in the region $[-3, 3]$ and the true stimulus being at 0.

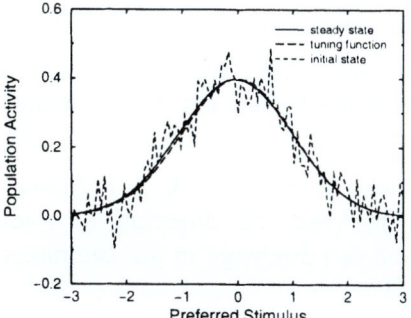
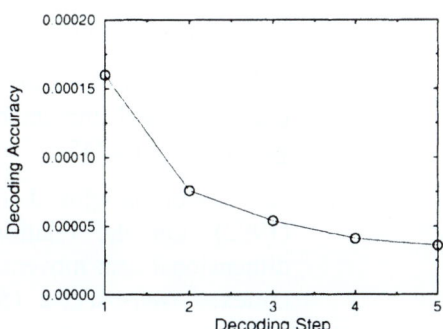

Figure 17.1. (a) The typical states of recurrent network before and after decoding. The tuning function is also shown in order to compare its shape with that of the steady state (being proportionally scaled). (b) The variance of decoding error decreases with steps. The learning rates at each step are $\eta_1 = 100$, $\eta_2 = 200$, $\eta_3 = 300$ and $\eta_4 = 400$. Other parameters are $a = 1$, $\mu = 1$, $\sigma^2 = 0.0004$, $\epsilon = 0.1$, and $T = 20$, where T is the time duration of network relaxation.

5. Conclusion and Discussion

In summary, we have discussed how to implement a hierarchical Bayesian decoding in a neural circuit. In the first step when there is no prior knowledge, the network implements MLI, whose estimation is subsequently used to form the prior distribution of stimulus for consecutive decoding. In the second step, the network implements MAP, whose estimation forms the prior for the third step decoding, and this process is repeated.

Line attractor and Hebbian learning are two critical elements to implement Bayesian decoding. The former enables the network to do template-matching by using the tuning function, being as same as MLI and MAP. The latter provides a mechanism that conveys the prior knowledge obtained from the first step to the second one. Though the result may quantitatively depend on the formulation of models, it is reasonable to believe it is qualitatively true, as both Hebbian learning and line attractor are both biologically plausible. We expect that the essential idea of Bayesian inference of using prior knowledge to enhance consecutive decoding is used in the information processing of the brain.

Acknowledgment

We are grateful for the reviewers' valuable comments.

References

[1] S. Amari (1977). Dynamics of pattern formation in laterial-inhibition type neural fields. *Biological Cybernetics*, **27**, 77-87.

[2] S. Deneve, P. E. Latham & A. Pouget (1999). Reading population codes: a neural implementation of ideal observers. *Nature Neuroscience*, **2**, 740-745.

[3] A. P. Georgopoulos, J. F. Kalaska, R. Caminiti & J. T. Massey (1982). On the relationship between the direction of two-dimensional arm movements and cell discharge in primate motor cortex. *J. Neurosci.*, **2**, 1527-1537.

[4] J. H. R. Maunsell & D. C. Van Essen (1983). Functional properties of neurons in middle temporal visual area of the macaque monkey. i. selectivity for stimulus direction, speed, and orientation. *J. Neurophysiology*, **49**, 1127-1147.

[5] M. A. Paradiso (1988). A theory for use of visual orientation information which exploits the columnar structure for stryate cortex. *Biological Cybernetics*, **58**, 35-49.

[6] A. Pouget & K. Zhang (1997). Statistically efficient estimation using cortical lateral connections. *NIPS*, **9**, 1997.

[7] A. Pouget, P. Dayan & R. Zemel (2000). Information processing with population codes. *Natural Neuroscience Review*, **1**, 125-132.

[8] S. Wu, H. Nakahara, N. Murata & S. Amari (2000). Population decoding based on an unfaithful model. *Advances in Neural Information Processing Systems*, **12**, 192-198.

[9] S. Wu, D. Chen & S. Amari (2000). Unfaithful population decoding. *Proc. IJCNN2000*, July, Como, Italy.

[10] S. Wu, H. Nakahara & S. Amari (2001). Population coding with correlation and an unfaithful model. *Neural Computation*, **13**, 775-797.

[11] S. Wu, S. Amari & H. Nakahara (2001). Population coding and decoding in a neural field: a computational study. *Neural Computation* (in press).

[12] S. Wu & S. Amari (2001). Neural implementation of Bayesian inference in population codes. *NIPS 14* (in press).

[13] K. Zhang, I. Ginzburg, B. McNaughton & T. Sejnowski (1998). Interpreting Neuronal Population Activity by Reconstruction: Unified Framework with Application to Hippocampal Place Cells. *J. Neurophysiol.*, **79**, 1017-1044.

Index

affine
    arithmetic, 2–5, 7, 16–18, 144–148
    interval, 3, 4
    interval arithmetic, 5
animation
    image-based, 176
    model-based, 176

Bayesian
    classifier, 157
    inference, 121
    population decoding, 200
Bernstein basis, 145, 146, 187, 188, 196
bivector, 42, 49, 52, 56
boundary representation (B-rep), 59, 60, 70
BUILD solid modeller, 71, 72

Clifford algebra, 43, 47
conformal
    group, 46
    inner product, 46
    invariance, 57
    model, 43, 51
    space, 46, 51, 55, 56
    transformation, 46, 49
conic
    absolute, 43
    flatness, 80
    sections, 81
constructive solid geometry (CSG), 59, 60, 70
conversion
    B-rep to CSG, 60
    Bernstein basis to power basis, 149
    CSG to B-rep, 60, 94
    power basis to Bernstein basis, 149
convex hull, 145, 149, 150
Cramer–Rao bound, 110
cross-validation, 123

determinant
    sign of, 97
distance to singularity, 192
dual vector, 53

entropy, relative, 30
exact geometric computation, 92, 95
expectation maximisation (EM) algorithm, 166, 167, 169
exterior product, 42

facial animation, 175
Fischer information matrix, 116
floating point filter, 93, 96

Gaussian process, 29–34, 121, 124, 128
generative topographic mapping (GTM), 166, 167, 169
geometric
    algebra, 43, 47, 48, 56
    primitive, 51
Gram matrix, 34
greatest common divisor (GCD), 186, 191

healing, 87
Hebbian learning, 204, 205
hidden Markov model (HMM), 177, 178

implicit
    curve, 15
    surface, 15
implicit linear interval estimation (ILIE), 20, 23
interval arithmetic, 7, 16, 17, 144–147

join, 56

kernel, 133, 134
Kullback-Leibler
    distance, 30
    divergence, 34

Levin's method, 62, 64, 80
Lie
    algebra, 49
    group, 49
localization error, 108, 110, 112, 113, 115

Markov chain Monte Carlo (MCMC), 35, 123
matrix

companion, 189
ill-conditioned, 98
norms, 156
resultant, 186, 190, 191
maximum *a posteriori* (MAP), 201, 202, 204, 205
maximum likelihood inference (MLI), 201, 202, 204, 205
medial axis, 94, 101
Mercer's theorem, 134
modified enumeration algorithm, 22, 23
Monte Carlo sampling, 124, 128
multivector, 47, 52, 53, 56

nearest neighbor classifier, 156, 157

outer product, 42

pattern classification, 139, 155
pencil, 62
Pluecker
    condition, 42
    coordinates, 42
point distribution model (PDM), 176
pole, 44, 45, 82, 84, 85
polynomial
    Bernstein, 188, 189
    root isolation, 94, 96
    Wilkinson, 193
population coding, 199, 200
principal component analysis (PCA), 176, 177
projective geometry, 41-43

quadric
    ruled, 62
    surface, 61, 62
quadric surfaces
    curve of intersection, 59, 60, 64, 80

range arithmetic, 96
regression, 136, 139, 155
regularization
    machine, 132
    network, 131, 133, 136, 137
reproducing kernel Hilbert space (RKHS), 132, 133, 135, 136, 138, 139
resultant, 186
resultant matrix
    basis transformation, 195-197
    condition number, 192
ROMULUS solid modeller, 71
rotor, 49-51

seam, 83, 85
singular value decomposition (SVD), 98, 192
stereographic projection, 43, 44
structure from motion, 108, 110
Sturm sequence, 95, 99
subdivision
    adaptive, 16
    surface, 70
    uniform, 16
support vector machine (SVM), 131-133, 137, 139
surface inversion, 81

timelike, 52
tracking error, 108, 112, 115
trivector, 52, 53, 55, 56

MIX
Papier aus verantwortungsvollen Quellen
Paper from responsible sources
FSC® C105338

If you have any concerns about our products,
you can contact us on
ProductSafety@springernature.com

In case Publisher is established outside the EU,
the EU authorized representative is:
**Springer Nature Customer Service Center GmbH
Europaplatz 3, 69115 Heidelberg, Germany**

Printed by Libri Plureos GmbH
in Hamburg, Germany